Contemporary Issues: Science, Africa, and More

Joshua Spencer

CONTEMPORARY ISSUES:

SCIENCE, AFRICA, AND MORE

JOSHUA SPENCER

CONTENTS

DEDICATION

This book, *Contemporary Issues: Science, Africa, and More* is dedicated to my three children, Kadisha Spencer, Kaliese Spencer Carter, and Michael Spencer, my late foster parents, Mr. Jim (James) Thompson and Mrs. Elethia Thompson (Operators of Bar-B-Barn Hotel & Restaurant, Negril, Jamaica), as well as to all the students of our global village – from junior kindergarten through to university.

FOREWORD

Contemporary Issues: Science, Africa, and More is a compilation of original essays on modern issues, some of which are of a moral nature and are highly debatable topics. These topics are not merely relevant to the twenty-first (21st) century but are matters that affect or will impact the wellbeing of all, positively or negatively. These issues, if embarked upon, endorsed, or implemented by the political authorities of the day, could lead to radical, social, economic, and political changes in the ways the lives of citizens of the globe are conducted today or in the near future. Some of these essays were written by the author as papers to satisfy degree requirements. Others were written out of the sheer interests of the author who is a teacher and author of six published books, including this one.

A good portion of this book attends to very sensitive issues pertaining to biotechnology such as human cloning and genetic engineering. It also comprises such areas as Affirmative Action, one's right to abortion, one's right to sexual choice(s), questions relating to whether Dr. Martin Luther King, Jr. was a hero or not. It also has other similar but controversial issues. In addition, there are topics included in *Contemporary Issues: Science, Africa, and More* relating to matters of ethnicity, culture, and development, in particular as they impact Africa.

In essence, there is something in this book that will command the interest of every citizen on planet Earth. There are numerous benefits to be derived by reading or studying from this text. Those studying Philosophy or Law will augment their own skills of debate and their organizing and presentation of arguments. Students of History and Social Studies, as well as students of African studies, will benefit immensely from this text and will acquire a new outlook on culture and its usefulness in Africa. Students of Science will sharpen their knowledge and skills in the areas of biotechnology in the areas such as Molecular/Cellular Biology, and in particular in the understanding of biological concepts such as cloning, genetic engineering, deoxyribonucleic acid (DNA), stem cells, embryonic stem cells, and much more.

In addition to the above facts, visible minority groups in North America and Europe, such as Jews, blacks, individuals of other ethnic extractions, gays, and lesbians, may all find specific areas of the text that will elicit their interests, engagements, and most probably, will garner, on their part, some optimistic outlook after reading it. Canadians will all acquire a more rooted comprehension of their Canadian rights as outlined by their Constitution.

Each topic discussed in the book ends with a *Test Your Learning* segment. Each *Test Your Learning* segment comprises five questions. These questions assist you in assimilating the material in the reading. If, after reading the section of the text, one cannot fully and completely respond to the questions, the material of that section should be reread.

ACKNOWLEDGEMENT

Although the essays in *Contemporary Issues: Science, Africa, and More* were written over a period of approximately one decade and are all my original work, I find it pertinent to advise that my proclivity to do all types of writings (poems, essays, short stories, etc.) were all encouraged and made possible by a number of important individuals and significant situations in my life.

First, I must thank my foster-parents, Mrs. Elethia Thompson and Mr. Jim (James) Thompson of Jamaica, for providing me with the opportunity to attend Manning's High School and later, Sam Sharpe Teachers' College where I was given a solid foundation in my educational pursuit. At the latter institution, I majored in Secondary Science and Physical Education.

Secondly, during the 1980s, as a young man, I read a couple of my poems at a youth gathering of which former prime minister of Jamaica, The Honourable PJ Patterson, Q.C., Attorney-At-Law was in attendance. After the function, he walked up to me to commend me for the excellent poetry, which unknowing to him, had given me great encouragement since that day to continue my quest of writing.

Thirdly, I must acknowledge the Late Dr. Norman Buchanan, also a Jamaican who had also played a stellar role in my personal growth and development. (A biography of Dr. Norman Buchanan, *A Quest for a Dream – A Life Committed* to Progress) was released on August 17, 2007 on the life of this great individual. The book is authored by his mother, educator, social worker, and poet, Joyce Buchanan.)

As a student of York University, Toronto, in the nineteen nineties, I had also been groomed by some very scholarly professors. Among them were professors who taught me in the areas of Psychology (my major at the university), African Studies – Professor P. Idahosa, Canadian Studies – Professor Dan Azoulay, Philosophy of Law, Spanish (Advance and Intermediate), Science, Statistics – Professor A. Herzberg, who incidentally was the author of the textbook we used, Principles of Statistics by Paul A. Herzberg, Social Science, Professor Harold Kaplan, and many more. I thank them all for all their efforts and generosity in making me the successful, knowledgeable, and confident writer and person who I am today.

In addition to the foregoing, I must name Mr. Carlton Grant, Mr. Glendon Lawrence, Ms. Grace Lawrence, Dr. Hixwell Douglas, Mr. Dudley Jennings, Mr. Oliver Nelson, Mr. Arthur Gilling, and Mr. Taiwo Osuntoyinbo, all college friends except the last, who was a colleague when I worked with CIBC Online Banking. These individuals have all unrelentingly, over the years, encouraged me to have what they considered my very scholarly work published.

I also must thank my former neighbour, Mr. Renato Punu (without whom, I would have died some time ago. He rushed me to hospital when I needed emergency surgery on February 17, 2007). He was also instrumental in my motivation and proclivity for writing.

Finally, I must acknowledge Dr. Simon Clarke, Dr. Cecile Walden, Dr. Millard Lowe, et al, all of whom played salient roles in my educational pursuit during my three-year stint at the Sam Sharpe Teachers' Training College, in Jamaica.

AUTHOR'S NOTE

The themes in *Contemporary Issues: Science, Africa, and More* are many and varied. However, the author's goal is to stimulate his readers' thoughts and keep the discussions on controversial, contemporary issues alive and current. It is only in so doing that there can be any certainty that we, the ordinary people, may influence future policies on matters that are sure to have a tremendous, enduring impact on our lives.

In addition to the above, it is the author's goal that *Contemporary Issues: Science, Africa, and More* will serve as a useful "resource package" for teachers and parents. It is hoped that it will provide them not only with convenience for review and discussions with their students and children, respectively, but will elicit more confidence, where it is deficient due to little or no exposure, knowledge, or experience with the subject matter discussed in this book. This will more likely guarantee that children of parents and students will have the opportunity to be introduced to the materials by their parents or teachers, topic/issues that may be perceived as somewhat challenging at times.

Having said the forgoing, I must advise that I am a mere trained teacher of Science. I am not an expert in the areas discussed. Even though I had taken great care in reviewing and presenting the subject matter I write in this text, I am not claiming to be a scientist, nor for that matter, an expert in any of the areas related to Science that I have discussed in *Contemporary Issues: Science, Africa, and More*. The information presented, therefore, represents a simple, rudimentary, workable guide. One that can be used as a foundation for deeper research and discussions in the areas articulated. Accordingly, it would be prudent, as well as recommended, that the candidates and pupils reading for examination purposes, especially at the community college level and advanced levels, etc., seek corroboration on the facts given here, from materials established and authored by known experts in their fields of specialization.

Contemporary Issues: Science, Africa, and More is therefore useful not only for those who want to read for personal awareness, development, and growth, but is mandatory for all teachers (especially those who were not trained in Science), parents, local community and national leaders, students and the political directorates of every nation on the globe. Its simple language facilitates easy mastery of otherwise challenging concepts. It is, indeed, a great foundation for pupils of Science, Philosophy, students interested in debates, curious, inquisitive individuals and others, who require some assistance in contemporary issues and who believe in maintaining and grooming a sharp wit.

ESSAY ONE

SHOULD HUMAN CLONING BE ALLOWED?

Since the cloning of the sheep, Dolly in 1997 by scientists, Dr. Ian Wilmut and Dr. Keith Campbell of the Roslin Institute near Edinburgh, there have been much discussions and debates as to whether human beings should be cloned as well. In this essay, I will share some of the arguments that proponents of human cloning may proffer as well as those that may be furnished by others who are in opposition to human cloning. To give a deeper sense and comprehension of the arguments that may be put forward for, or against, human cloning, and to establish and fortify my readers with the requisite knowledge to arrive at an informed position on the issues, I will define the term, cloning, and other science based terminologies as well as identify the different types of cloning and their intended significance as may be purported by the scientific community. My intention, in this paper, is simply and merely, to provide probable information from both sides of the debating fence on these issues. In so doing, my readers, you will be able to come to your own, informed conclusion, thereafter.

WHAT IS CLONING?

Cloning is, in simple terms, the transferring and implanting of the *Deoxyribonucleic Acid* (DNA) from the *somatic cell* (*body cell*) or any other cell of a living organism into the female's egg (*ovum*). Before this transferred DNA is implanted into the female's egg, the nucleus of the egg is removed. The nucleus of each cell, whether of female eggs, regular body cells (*somatic cells*) or *sperm gametes* (*spermatozoa*), contains DNA. Once the transferred DNA is implanted into the female's *enucleated* egg (an egg whose nucleus has been removed), an electric shock is applied by the scientist. As a consequence, the egg exhibits behaviour similar to a fertilized egg (i.e. it behaves like an egg that has been fused to a sperm through normal copulation in the sexual act) and starts its development into a new organism (the *blastocyst* first, then the *embryo*) but with the same DNA components (or very close to the same) as the organism whose original DNA had been transferred into the female's egg (*ovum*).

MORE ON DNA

Deoxyribonucleic Acid (DNA) is a very useful substance and important substance (*molecule*) located in all living cells. A *Molecule* is a substance in which two or more *atoms* are combined. *Atoms* are the basic unit of matter, consisting of positively charged protons, negatively charged electrons and neutrons, having no charges. The number of protons in an atom is equal to the number of electrons. A molecule may contain the atoms of only a single element, such as oxygen (O_2) only or the molecule may contain atoms from different elements that are combined, such as in water (H_2O). In the case of a *molecule* of oxygen, two atoms of oxygen are combined. As a matter of fact, oxygen atoms always form molecules and do not exist as single atoms. In the case of water, two *atoms* of hydrogen combine with one *atom* of oxygen to form a water *molecule*.

Deoxyribonucleic Acid (DNA) comprises information that is used in the *building up* (*anabolic*) processes and the *breaking down* (*catabolic*) processes of the cell (together these processes are known as *metabolism* or *metabolic processes*), required in the everyday chemical reactions of the cells of living organisms. It also influences most, if not all, of our characteristics. It is often referred to as the blueprint of living organisms, plants, and animals. From a scientific perspective, animals also include human beings and lower-class animals, such as cows, pigs, and so on.

DNA is so described, as it aids cells to develop and function together, to create a complete functional body or being. DNA also controls such characteristics as eye and hair colour, and is believed to play an overwhelming, salient impact on one's intelligence.

The information that DNA comprises is transferred from parents to offspring during the normal fertilization of the female's *ovum* (egg). In other words, the fusing of the sperm with the *ovum* or, in the case of *asexual reproduction* that takes place in some plants and animals and also in the case of cloning. There is no consensus, however, on what level or proportion of our characteristics that is actually as a result of inheritance and is determined by our DNA (nature) or is as a result of our associations and influence by our surroundings or environment (nurture).

Genes are usually made up of short lengths of DNA, and modern gene technology, such as genetic engineering is able to make changes at the level of individual genes.

TYPES OF CLONING—THEIR PURPOSE

In most scientific parlance, there are two main ways to name or describe cloning that may take place as it relates to human beings, though other descriptors may sometimes be used interchangeably for either type. There is the potential for cloning which will eventually result in a fully blown, cloned adult from the embryonic stage if allowed to develop. This type of cloning is usually referred to as *Reproductive Cloning*. In this case the cloning is for the purpose of producing a second human from embryo to an adult that has been cloned to share the same DNA as its original subject. In the second stead, there is cloning whose intent, according to these scientists who are involved in such innovation, is to serve as a basis for research and to cure or prevent diseases such as Alzheimer's, Parkinson's, cancer, etc. This kind of cloning is usually described as *Therapeutic Cloning*. During *Therapeutic Cloning*, DNA from an adult cell may be placed inside an egg to develop useful stem cells but not with the intention to produce young ones to grow into adults. This kind of human cloning, the scientific argument goes, would provide potential patients with the capacity to have their own biological material (genetic material) to be employed for future use of damaged cells, if the need arises. This is a secure means of ensuring that patients will not have to face the problem of organ or tissue rejections during surgery if an organ or tissue donor cannot be found with the proper match. It must be noted however, that even in the latter type of cloning, if the scientist does not manipulate and abort the reaction, an embryo will also be formed which has the potential, as in the first case described above, to grow into a full blown adult. This possibility has led to some opponents of human cloning to argue that there is no distinction between *Reproductive Cloning* and *Therapeutic Cloning* and that in fact scientists have simply created these biological terminologies for the sheer purpose of misleading and deceiving the unsuspecting public.

WHAT ARE STEM CELLS?

Stem cells are primal cells that are found in all multicellular organisms (i.e. organisms that are made of more than one cell, e.g. human beings). They possess the capacity to renew or make copies of themselves through mitotic cell division (i.e. the process in which a cell duplicates its chromosomes to generate two identical cells. (A chromosome is a single, large strand of DNA). Stem cells can also *differentiate* in a diverse range of *specialized cell* types. It is for the latter reason that scientists in the areas of *Therapeutic Cloning* research are so vocal and vigorous in their intent to persuade policymakers to support and fund this kind of cloning. Put another way, *embryonic stem cells* are *undifferentiated cells* that have the capacity to evolve into any of the *specialized cells* in the body of which they are two hundred twenty (220), such as blood cells, muscle cells, nerve cells, etc. It is hoped that in this way, they will have the necessary and *specialized cells* to replace or repair damaged cells in the bodies of human beings. The Scientists' goal, therefore, is to collect these *embryonic stem cells* before they *differentiate* and then mould them into becoming various types of *specialized cells* in the laboratory (lab). It may interest my readers to note that stem cell research emanated out of the findings of two Canadian scientists, Dr. Ernest A. McCulloch and Dr. James E. Till in the 1960s.

SOME CLARIFICATIONS ON UNDIFFERENTIATED CELLS, DIFFERENTIATED CELLS, AND SPECIALIZED CELLS

The term *undifferentiated cell* is used to describe a cell that has not yet taken on any specific form or function. That is, an *undifferentiated cell* has the potential to be any of the cells in a living organism. It can be made to be a nerve cell or a muscle cell or any other cell that the scientists may need to perform a potential therapeutic function. For example, the need may exist to use *undifferentiated cells* from *embryonic stem cells* to replace nerves that may be damaged in say, patients suffering from Alzheimer's. A *differentiated cell* is a cell that has a specific form and function. That is to say, a *differentiated cell* is a *specialized cell*. Example, it may be a liver cell, a muscle cell, a bone cell, etc. Put another way, the *differentiated cell* which can also be described as a *specialized cell* is an end in itself. It is a finished tissue or organ with a specific form and function. The *undifferentiated cells* found in embryos, may facilitate a process that may lead to several of any needed end(s) in terms of organs and/or tissues. These *undifferentiated cells* can be "bribed" by the scientist to develop into any of a required number of organs or tissues at any given time according to the need of the scientist, Medicine, or the patient.

SOME ARGUMENTS THAT PROPONENTS OF HUMAN CLONING MAY OFFER

Human cloning can be used to promote the wellbeing of a people. It can lead to a more quality and satisfying life. For example, if *Therapeutic Cloning* is allowed to have its way, may be the argument proffered by its proponents, one needs not worry about dying from some of the scourges and maladies that today permeate us and are responsible for the ailments from which many suffer and even die. Illnesses such as cancer and Alzheimer's would soon be things of the past. In the case of *Reproductive Cloning* women and men who are infertile will now get the opportunity to produce children related to them genetically and accordingly, will not have to settle for children through adoption, children having no biological link or connection to them as parents. Children, who, may or may not, be from eggs and or sperms of individuals who have the propensity to be evil and do evil things. This, in itself, leads to greater wellbeing and contentment on the part of these infertile couples. As a significant number of people (10%-15 %) throughout the world, is infertile, allowing human beings to clone themselves for the purpose of creating offspring, the argument may be, is playing an important and salient role in the lives of a significant number of the society. This will also alleviate psychological problems on the part of those affected, may be the added, supporting point proffered by this group. As a consequence, there may be a better mentally prepared workforce which would indirectly lead to augmented production on the part of the society. This, in turn, has the potential to create a more vibrant and prosperous economy for all.

There are still other reasons that proponents of human cloning may provide to support the use of this type of biotechnology. Some scientists may argue that challenges such as obesity and aging may be history if this kind of human cloning is allowed to be developed, and fine tuned, in particular, *Therapeutic Cloning*. Also, defects and diseases such as heart problems may now be able to be studied from this research and accordingly, will significantly reduce or completely eradicate these maladies. Expensive plastic surgery may no longer become necessary if the research team can identify genes or lack thereof, that cause or inhibit breast growth, for example. As breast implants may not be deficient in side effects, *Therapeutic Cloning* may be a good, if not the only practical, alternative in these situations. Other positives that may be offered in support for human cloning is that it may aid in alleviating kidney problems, eliminate, or alleviate the probability of Down's Syndrome, liver failure, among other serious diseases. Some may even hope, and argue, that the technology may lead to a better understanding of HIV/AIDS infested cells which could ultimately lead to either reducing the lethal power of the HIV/AIDS or to its complete demise.

SOME ARGUMENTS THAT OPPONENTS OF HUMAN CLONING MAY OFFER

Those who are against the human cloning biotechnology may present several arguments against it. It may be argued that human cloning is morally wrong as it involves the creation of life, then the turning around to destroy it. Put another way, human cloning may be considered by such persons as tantamount to committing murder and is no different from say, acts such as abortion and lynching. The point may also be made that the embryos which are needed to carry out the human cloning is a living organism. They may postulate that the embryos are no different from the potential adults and carry and possess all the genetic material blueprints that the full-grown adult will have, if allowed to develop to that stage. Accordingly, it is unfair, the argument may be, to make decisions on behalf of this living organism without him or her having any say or contribution in the arrival of the decision and how his or her future will be impacted as a result of such an important action. The point may also be made that all the rights of human beings need to be protected, irrespective of age, economic status, creed, and so on.

Apart from the reasons furnished above in opposition to human cloning, there may yet be other arguments in stifling and denying support to the human cloning technology. Some may present the point that human cloning not only will serve to dehumanize and commoditize individuals but also it strongly has the potential to augment the chasm between peoples of nations worldwide and even within their own, shared jurisdictions. The point may also be made that in allowing this technology to have its way women are simply being exploited in this regard. There is, for example, the point may also be made, the potential to totally erase a certain "genetic pool" of people while promoting the existence of others. In this case, the ramifi-cations are not only one that may cause personal, psychological, and political upheavals on the part of millions of peoples worldwide but could result in a negative impact on the economies of nations and how they relate with others internally and externally on the globe. This could happen if scientists choose to think along the lines of creating and programming everyone to be of a selected intelligence profile, to the detriment of others. Opponents may also argue that due to this artificial design, the necessary workforce of various hierarchal levels, necessary for the diverse work force of individual economies, may become nonexistent and significantly compromised resulting in the potential collapse of most economies. Others in opposition to human cloning may stipulate that the technology is too costly and inefficient. This is because the process of human cloning involves the destruction of several embryos and several time-consuming trials prior to succeeding in the creation of a single clone. It may also be argued against on the basis that scientists cannot arrive at a clear, unambiguous consensus as to the safety of individuals who would be created from this kind of technology.

Human cloning, while it may have the potential to be useful medically/ therapeutically and otherwise, equally has the capacity to be destructive. This perception has caused its opponents to be averse to its use. For example, political powers and authorities could use this technology to breed droves upon droves of evil soldiers, etc., equally as it could be utilized to produce, upright intelligent homo sapiens. The ramifications of such possibilities do not need elaboration. In a way, any game can or may be played, the opponents may echo in concern. One cannot leave such an important technology up to chance, may be the suggested viewpoint. On this basis, therefore, its critics will argue most vigorously and vociferously for its demise.

Finally, human cloning has the proclivity to marginalize intimacy and destroy the conventional family structure. The fact that human cloning makes traditional sexual intercourse theoretically and scientifically redundant creates the potential to eliminate sexual relations among homo sapiens to which we have grown accustomed. Secondly, the conventional mode of the family structure if at best, not challenged in such a process, is prone to become chaotic, confused, and confusing. How will one, for example, describe the relationship that obtains as a consequence of human cloning? Is one a "mother", the other a "child" or are "mother" and "child" mere siblings or in some cases, a grandparent/grand child? Who will determine who is parent, child, grandmother or whether the two are related or not? In such reproductive dilemmas, will there be new laws required to prevent situation as "the new incest" from occurring without any potential for legal, binding sanctions?

I have attempted to shed some light on human cloning and cloning as a whole. Several possible reasons that may be proffered for its support and application have been discussed. In the same breath, several points that may be brought to the fore against the technology have also been presented. It is now up to you, my readers, to take your position. I would humbly advise that you continue to read as much as possible as you can on the subject. Further, I implore you to continue the discussions, with friends, read science journals, and attend fora (forums) organized to this end, with a view to have a broader and clearer comprehension of the implications of this new technology. It's your future and that of your children that will surely be impacted!

TEST YOUR LEARNING

---

(1) Using information from the essay, additional resources such as your local and/or school library, define the term human cloning.

(2) Would you say there are differences between *Reproductive Cloning* and *Therapeutic Cloning*? Explain.

(3) Using information from the essay, additional resources such as your local and/or school library or by interviewing an expert, explain why a Therapeutic Technology Expert/Researcher

may be more prone to do her/his research with embryonic stem cells than with adult stem cells?

(4) At the outset of the reading, the author states that he is merely presenting both sides of the issue of human cloning as proponents and opponents of human cloning might have. (a) Do you agree that the author is merely presenting both sides of the issue? (b) Does he present any biases? (c) Are there additional arguments that he could have put forward for both sides? (d) Are there arguments that he puts forward on both sides that are irrelevant and do not provide supporting detail? Explain.

(5) Do you believe scientists should be allowed, and given public funding to do research to develop and finetune human cloning? Elaborate.

ESSAY TWO

IS GENETIC ENGINEERING HARMFUL OR HARMLESS?

During the dusk of the twentieth(20th)century and the approaching dawn of the twenty-first (21st), and in particular since the successful cloning of the sheep, Dolly in 1997 by scientists, there have been much commotions, discussions, researches, investigations and debates on the controversial but contemporary advances in the areas of biological and agricultural technologies. Much has been made of Genetic Human Cloning as discussed in the previous section of this book but there have been an equal flurry and controversy on the matter of genetic engineering in general as it relates both to human beings, other animals, and plants.

In this essay, I will share some of the arguments that proponents of genetic engineering may present as well as those that may be argued by others who are in opposition to genetic engineering. To give a deeper sense and comprehension of the arguments that may be put forward for, or against, genetic engineering, and to establish and fortify my readers with the requisite knowledge to arrive at an informed position on the issue, I will define the term genetic engineering, explain and outline the two main types of genetic engineering and other science based terminologies, as well as identify some of the different ways that this new technology is employed. I will also highlight their intended significance as may be purported by the scientific community. My intention, in this paper, is simply and merely, to provide probable information from both sides of the debating boundaries on these issues, and to allow my readers to come to their own informed conclusion, thereafter. Note the information here is treated and presented mostly in a "lay person" fashion to facilitate easy learning and comprehension.

In prelude to defining and explaining the relevant terminologies and concepts as they aid in the comprehension of Genetic Engineering, its relevance or lack thereof, it may be prudent to make a point or two at this juncture. Firstly, the point must be made that it probably would be naïve on the part of us all, whether we support the new technology or not, if we fail to understand that economic consideration, in most cases, is a salient determinant as to whether or not the new technologies will live or die! The advancement, or not, of most technologies will upon close analysis be decided on its economic impact even as other factors such as morality and health risks are considered. In the case of genetic engineering, the technology, according to the technocrats, promises an ocean of wealth to the practitioners due to relative augmented, qualitative, and quantitative anticipated proliferations in overall agricultural enhancement, for example. Put another way, private capital (read business) is bound to be, in most cases, a proponent for the genetic engineering technology as it augers well for augmented profits on their part even if its overall effects may be significantly negative in such areas as the environment, for example, or general health risks. In considering the pros and cons for the biotechnology of genetic engineering, it may be crucial for you, the reader, to consider all the facts available, including the above point mentioned and those to follow.

WHAT IS GENETIC ENGINEERING?

Genetic engineering is the manipulation of a living organism's genes by the use of technological interventions. In so doing, the genetic engineering scientists use the technology to change the genetic expression, nature, and make-up of the living organism.

Note that this living organism can be an animal, including human beings or a plant. This manipulation of the genes of the organisms leads to the creation of an organism with a desired or needed trait(s) and the removal of the unwanted ones. Put another way, genetic engineering is a structured group of technologies employed to change the genetic makeup of cells and to transfer genes across living organisms (plants and animals, including human beings) irrespective of their species. That is to say, genes can be transferred from say, a cow to a goat, and so on, to have a desired outcome or trait. It is important for you, my reader, to note that cloning, including human cloning (already discussed in the first essay of this book), is a component of genetic engineering.

MORE ON GENETIC ENGINEERING

Here are some further elaborations and examples that will certainly leave you with a fuller and more complete comprehension of genetic engineering.

Genetic engineering may also be described as the alteration of the genetic code of the living organism by artificial means and differs from the conventional *selective breeding*. The conventional or traditional selective breeding, which is synonymous with the term, artificial selection, involves human beings selecting two specific animals for copulation (sexual intercourse) in order to acquire specific traits or characteristics in the newborn animal. As a child, I witnessed this a lot growing up in a small district known as New Mills. Owners of cows, dogs, or other animals would seek certain, chosen animals of their neighbours with which to breed. In effect and reality what these neighbours were doing was essentially choosing two animals to mate that will very likely result in high quality, beneficial phototypic traits in the offspring of these animals derived from parents possessing these very beneficial traits or characteristics. Genetic engineering does the contrary. It manipulates genes of a given organism to derive a specific outcome or trait.

SOME EXAMPLES OF GENETIC ENGINEERING AT WORK CAN BE SEEN IN THE FOLLOWING CASES:

(1) The injection (insertion) of the human genes into sheep. This results in the secretion of a very useful chemical substance (alpha-1 antitrypsin) in their milk which is considered useful in the treatment of many types of lung diseases.

(2) The extraction of the gene that programs poison in the tail of a scorpion and combining or mixing it with cabbage. These genetically modified cabbages, as a consequence produce a special type of insecticide in their saps that serves as a means of self-defence for the cabbages against being destroyed by

"predators" such as caterpillars. The result is that caterpillars that use the cabbages as their meal are killed!

(3 Scientists have also conducted and created some kind of "weird" activities by the application of genetic engineering. For example, it was said, and believed to be true by some, that scientists had created (manufactured? a four-legged, no-winged chicken. There are also reports that scientists had also made a goat genetically, with spider genes that produced "silk" in its milk.

Genetic engineering, in all seriousness, however, works because of the flexibility of genes to different species or organisms. It would appear that genes are not species-specific or prejudice to development, growth or evolution in terms of where they will proliferate and make their presence be felt. Human genes, for example, will continue its impact and trait, irrespective of whether it is transferred to an ape, a mouse, a snake, a cow, or the hibiscus flower! Frog genes conduct or express their impact in rice, and tree genes act accordingly in fruits. There is absolutely no boundary theoretically, for the capacity and potential of genetic engineering. This, for all the proponents of the biotechnology, is very exciting news!

SOME BASICS ON HUMAN REPRODUCTION

There are several debates and discussions on the issue of when a human life begins. For some pro-life advocates including clergies and others, human life begins the moment the *spermatozoan* (*sperm cell*) fertilizes (fuses) with the egg (*ovum*) of the female. On this premise, therefore, is the origin for or against, the argument as to whether or not genetic engineering, in particular human cloning, is moral or immoral, should or should not be allowed. But what are the researched, investigated, and established scientific facts, as they pertain to reproduction in animals including homo sapiens (human beings)? I will disclose this in a moment after reviewing some additional, basic pieces of scientific facts to you.

It is common knowledge that human beings (and other living, multicellular organisms including animals and plants) comprise billions upon billions of living cells juxtaposed together. This juxtaposition of the cells, if you will, is comparable to the bricks or blocks that exist together in the construction of a building, your apartment or house, for example. These bricks (*cells*) combine in the human body to form *organs* and *tissues* such as muscles, kidneys, heart, and so on that have specific roles to perform in the maintenance and proliferation of life. The cells that make up the body are called *somatic cells*. This distinguishes them from the *reproductive cells*, namely the eggs (*ova*) of the female and the *spermatozoa* (sperm cells) of the male which are the *germline cells* that are responsible for linking the generations. It is now a common established, scientific fact that all life forms evolved from a single-cell organism approximately a billion years ago. It is, therefore, reasonable to conclude that life, in terms of the cell, is not a new beginning. Put another way, the cells, homo sapiens and other life forms have today, are ones that can be linked to the descendants of cells in existence from the original life form evolved on our planet over a billion years past. However, the discussion on the onset of life to determine, say, whether human cloning (*genetic engineering* of humans) is immoral, will not and cannot be based on the scientific history postulated. It will, more credibly and deductively, be based on the reproductive connection of cells as it relates to the formation of the embryo, to the foetus, to the baby. This takes us back to the question, "When does life begin?"

All human beings (homo sapiens) and other animals, as discussed in the foregoing, are made up of *somatic cells*. It is also a scientifically established fact that all *somatic cells* are related due to their common origin of the single cell that evolved over a billion years ago. During fertilization (the fusing of the sperm with the egg) a single cell is generated to form that which is known as the *zygote*. The *zygote* rapidly divides during the first week of fusion (*fertilization*) to form a ball of cells, not discernible by the naked eye, known as the *blastocyst*. It is sometimes also referred to as the *pre-implantation embryo*, a fact that speaks to the reality that it is not yet implanted in the *uterus* (womb) at this stage. It is interesting to note as well that scientists have also discovered that approximately <u>forty percent (40%)</u> of all *pre-implantation embryos* developed at the stage of normal sexual reproduction, does not get embedded in the wall of the *uterus* (womb) and are therefore destroyed as a consequence. It is also important to note that scientists have also established that at the *blastocyst* stage, there are no *somatic* (body) *cells* in existence or formed and there is no indication at this point either, for any remote signs of *differentiation* of the cells of this small, microscopic ball roaming around in the womb (*uterus*).

Finally, the following scientific facts should be noted by readers. The *embryo* is not implanted in the uterus (womb) until about fourteen (14) days after fertilization (the fusing of sperm with egg). Beginning at the time of implantation to the wall of the womb at, and subsequently to the fourteen-day period, the *embryo* begins to form what is described by scientists as *primitive streaks*. The *primitive streaks* formed can be one, two, three or more. These *primitive streaks* are a scientific indication that the *embryo cells* (*embryotic cells*) will form an individual homo sapien (human being). If there is only one, single *primitive streak* formed, the *embryo* will develop in a single, individual adult. If the *embryo* develops two *primitive streaks*, it grows into identical twins. Sometimes, and very rarely, the embryo develops two separate *primitive streaks*, but these streaks do not become completely separated. In such an eventuality, conjoined twins (Siamese twins) are formed. Even more amazing and even more rarely, there are cases in which two separately fertilized *ova* (eggs), adhere themselves together to develop a single *embryo* with two different types of cells. This results in a single homo sapien (human being) with some of the cells in their body being female from the original female *embryo* and some cells being male from the original male *embryo*. This is described scientifically, as tetragammetic chimera.

WHAT ARE GENES?

Genes are the chemical blueprints that determine an organism's characteristics or traits. The transferring of genes from one living organism to another, in effect, is the transfer of traits or characteristics from the "transferor" to the "transferee" The transferring of genes, and by extension, genetic engineering, makes possible the reality of organisms getting new combinations of genes and accordingly, new combinations or mixtures of traits or characteristics. These new characteristics (traits) do not occur in nature and cannot develop through natural means. As these new organisms are unique and not a product of natural creation, they may sometimes be referred to as *novel organisms*.

NOVEL ORGANISMS

Novel organisms, as described above, are developed through genetic engineering. They are organisms that would not have been able to be developed naturally or by nature. Nature, whilst it is able to produce new organisms with new gene mixtures, through sexual reproduction is significantly limited with the number of new combinations of offspring that may be reproduced. For example, if a genetic scientist needs to create a blue cow, it would be potentially very easy for her or him to do so. If the blue genes are available in nature, irrespective of the organisms in which it is found, the blue cow could be produced readily by applying the biotechnology. In a sense, most, if not, all organisms that are the creation of genetic engineering may, and could be correctly described, as novel organisms. They are not possible to be created by nature or naturally and are fundamentally different from all naturally created species of organisms.

SOME CLARIFICATIONS ON GENETIC CODE, GENETIC BREEDING, AND SELECTIVE BREEDING

Genetic code is the set of rules by which information encoded in *genetic material* (DNA and RNA (*Ribonucleic Acid*) sequence—*See the section on *Human Cloning*) is translated into proteins by living cells. *Genetic breeding* is basically breeding that arises from the manipulation of genes to produce an organism with a desired trait or character different and apart from breeding as in the case of sexual reproduction. Put another way, *genetic breeding* is *genetic engineering*. *Selective breeding*, as discussed before, in essence, it involves, to some extent, some of the same basic ideas of genetic breeding. When breeders and producers of animals and crops are interested in producing certain specific traits such as, resistance to disease, high crop yields, and so on, they select two organisms having the dominant genes in both organisms and for which they are interested in developing in the offspring. As both organisms have the trait as dominant, it will be expressed in the offspring thus proliferating the needed ancestor's traits.

At this point, prior to examining some of the arguments that may be proffered for, and against, genetic engineering, I will outline the two broad areas of genetic engineering. In so doing, it must be pointed out that genetic engineering is classified in these two categories based only on how it is applied. That is to say, its application is what determines its name. The two forms of genetic engineering are _somatic engineering_ and _germline engineering_.

SOMATIC ENGINEERING

One application of _genetic engineering_ incorporates and involves effecting changes to the genes in the cells of the body (soma) but not the _ovum_ (egg) or sperm cells (_spermatozoa_). In this application, the changes performed are not passed on to the children or offspring. This kind of _genetic engineering_ is known as _somatic genetic engineering_ or simply, _somatic engineering_. This kind of biotechnology is considered more socially acceptable by some and there are numerous clinical tests going on, even as I type this sentence.

GERMLINE ENGINEERING OR GENETIC GERMLINE ENGINEERING

The second type of _genetic engineering_ involves carrying out changes in the egg, sperm cell or early _embryo_. This application of _genetic engineering_ has significant impact not merely on future children and offspring of a wide group of species but for whole droves of future generations. It, in a sense, "flies the gate" for a whole gamut of biological travesty and the reconfiguration of homo sapiens (human beings) and other animals as well as plants. When the genetic engineering science is utilized in this sense, it is known as genetic germline engineering or simply, germline engineering. It is so described because female eggs (*ova*) and sperm cells (*spermatozoa*) are biologically the "germline" or "germinal" cells.

WHAT WAS THE HUMAN GENOME PROJECT (HGP)?

The *Human Genome Project* was an international research project used to discover and map all the 30,000 to 40,000 or more human genes, and to determine the exact sequence of the approximately three billion nitrogen base-pairs of homo sapiens' DNA.

WHAT IS MEANT BY MAPPING AND SEQUENCING THE DNA?

Mapping is the process of determining the positions of *genes* on a *chromosome* and the distance between them. A *gene* is a basic physical and functional biological structure or unit of an organism that is derived or inherited from parent organisms. The genes are made up of DNA. As alluded to earlier, some of the DNA (Deoxyribonucleic Acid of these genes serve to provide instructions to produce specific protein molecules. Others are involved in different functions as you gleaned earlier.

The sequencing identifies the order in which the basic chemical units (A's, T's, C's and G's of DNA appear. It is the order that the chemical units appear in the DNA that is the basis for all diversity, and determines not merely traits such as eye colour and hair texture, etc., but whether an organism is a human, a fruit fly, a frog or a giraffe.

HAS THE RESEARCH EMANATING FROM (HGP) ONLY FOCUSED ON HOMO SAPIENS (HUMAN BEINGS)?

Scientists also have worked on the *genome* of less complicated organisms such as yeast and the fruit-fly. Some years ago, the *genome* of the SARS virus was decoded. The knowledge of how the genomes of simpler, living organisms works, is expected to lead to a truer and fuller comprehension of the more complexed living things, such as homo sapiens (human beings.
In its research on human beings, the *Human Genome Project* used DNA taken from volunteers whose identities were withheld to ensure the privacy of these individuals.

ARE THERE POTENTIAL BENEFITS AND DANGERS OF THIS KNOWLEDGE?

Although DNA dictates everything a cell does, not all is clearly and wholly comprehensible of DNA. What is known is subtle eventualities, such as how the body responds to various chemicals (Example, alcohol and nicotine). Some believe that sometimes changes in the DNA may have the potential to lead to sickness, to diseases, and so scientists are still investigating these possibilities.

Decoding of the DNA was done as there was the expectation that this would lead to tremendous improvements in the diagnosis, prognosis, and treatment of diseases and illnesses, to create healthier crops and livestock, and to make improvements in criminal investigations. Some of the dangers could include the creation of designer babies (choosing traits of your child) and genetic discrimination, some may argue. (These will be mentioned later in the essay to follow).

HOW IS KNOWLEDGE OF THE HGP USEFUL?

Knowing the sequence of all the nitrogen base-pairs even if it does not lead to cures directly or treatments for illnesses, can aid tremendously in the treatment and approaches to new and current diseases. The sequence must be studied to determine how the information is translated into proteins. Proteins are the important factors to understanding diseases and other traits.

SOME ARGUMENTS THAT PROPONENTS OF GENETIC ENGINEERING MAY OFFER.

There are possibly several arguments that can or may be put forth in support of *genetic engineering.* The first argument that one could possibly make for *genetic engineering* is that it creates new avenues for the rapid development of crucial medicine and for providing human beings with a more enhanced and beneficial life and lifestyle. As the technology of *genetic engineering* improves and proliferates, it lends itself to deeper knowledge and skills in the handling and treatment of diseases and provides a fuller understanding of the molecular basis of health in relations to the roles of genes or their absence thereof. *Genetic engineering* is the saviour the world has been waiting for, its proponents may postulate in this regard. Secondly, its proponents may further argue that with the reservoir and inundation of information available to experts on genes and their roles in the cause of diseases or the suppressing of them, derived through the knowledge imbibed via the *Human Genome Project*, referenced above, the sky is the only limit in terms of the quality benefit humankind, animals, and plants may attain from the biotechnology of genetic engineering. They may also point out that genetic engineering will lead to faster care, diagnosis, and treatments for all, as well as earlier diagnosis and prognosis, greater numbers of peoples benefitting and fewer side effects in taking care of patients.

There are additional benefits to be derived from *genetic engineering*, its supporters may show. A possible and probable point that may be made in its support is that in cases where certain adult individuals have the propensity to pass on the gene for diseases such as cancer to their offspring, this could be quickly and easily prevented. The gene responsible for such an undesirable phenomenon could be easily isolated and destroyed through the biotechnology of genetic engineering. This lends to healthier children which in the long run leads to a healthier society. This in turn promotes the creation of a more productive economy with a more mentally and physically well-endowed class of individuals. Individuals may also be able to benefit from the technology to the extent that it may be able to eradicate certain addictions such as smoking and so on, as a result, as well.

The supporters of *genetic engineering* may possibly demonstrate other benefits from the scientific innovation of the technology. The argument could be made that some living organisms manufacture compounds that have inherent values that are of a therapeutic nature. For example, it is a known scientific fact that the majority of antibiotics available to our medical practitioners is made from microbes and there are several types of medicines available of which plants are their original sources. With genetic engineering and its continued development, not only will the supply of these needed products be available, but its output would be significantly strengthened and more likely, made more effective. Among some of the benefits that could arise from these plants are cures for diseases that have, to date, evaded the available medical treatments, may be the argument.

There are still other possible positive points to make for supporting the technology. One that may possibly be introduced in such a discussion, by its proponents, is its use in the area of gene therapy. Gene therapy utilizes genes to treat diseases. So, instead of say, giving regular doses of injections to treat a disease, gene therapy can be used to provide a replacement gene that will give the required outcome or simply isolate the defective gene, etc. Some may even argue that it may be possible to use gene therapy to eliminate or prevent diseases such as cancer. Genetic engineering may also be viewed as a positive phenomenon in that, in addition to the above, it is possibly useful in situations such as in the case of patients in need of organ transplants, among other uses.

Genetic engineering is not only harmless but an imperative and useful technology, some may argue quite vociferously. The point may be made that at the rate at which the Earth's population is expanding, traditional agriculture may not be able to meet the global food demand. With genetic engineering with its component procedures of genetically modifying crops, not only will food production be enhanced quantitatively, but it will do so qualitatively and nutritionally as well. Genetic engineering also offers the opportunity to create crops, through gene manipulations, that will result in plants that are immune to pests and viruses. Consequently, they will need little, if any, pesticides and herbicides. As, unlike the traditional crops, genetically modified crops will need little or no pesticides and herbicides, this serves as a boost to our environment and to protect it. In addition, those supporting the technology may also argue that many of the applications that are currently applied through the biotechnology to our crops are not new. That they are, and were, already practised in the past. Genetic engineering simply enhances and accelerates the conventional processes already being practised for ages, may be the argument! It may possibly be queried, "Then why should this technology be prevented?"

Finally, there may be support for the technology as it helps in the preservation of crops while being transferred from field to market. The technology allegedly can be used to prevent certain crops from getting soft or spoiled through long, rugged transportation from point A to Z, for instant. With the current technology, the geneticist may also be able to effect certain modification of the crops' genes to create these desired results. It, accordingly, would be desirable as it offers great economic benefits to farmers and directly to the economy as a whole.

SOME ARGUMENTS THAT OPPONENTS OF GENETIC ENGINEERING MAY PROFFER

There are several arguments that opponents of genetic engineering may present to demonstrate its harmful effects and undesirability. Firstly, the point may be made that genetic engineering will be used, especially in capitalist-oriented economies, chiefly, and above everything else, to create wealth at all cost. In this stead, factors such as the health of the environment, ecosystems, animals, and plants will be insignificant and secondary to everything else. As a consequence, these thinkers may argue, that it is not only important that the technology be stopped immediately before it's too late, but it is imperative that it be eradicated. Some may argue as well, that the methodology involved in modifying plant genes, will, inevitably, lead to the degradation not only of the environment but the specific plants themselves. To effect the genetic modification in plants, beads of gold are shot (blasted) from the barrel of a "gun" at approximately one thousand miles per hour (1000 Miles/Hour). This is what is claimed by scientists who are opposed to the method used to effect the genetic modification. It leaves these plants with a number of tissues that are targeted to undergo genetic modification, seriously damaged. Accordingly, only a few survive this "torture" of assault of gold beads. Just a tiny fraction of the plants will survive and display the required traits being attempted to be transmitted. Another possible concern by the opponents is that the rapid proliferation of genetically modified plants, especially in North America and Europe poses a great and dire threat to the natural ecosystem and may consequently result in serious calamity in the future. The extent of which, homo sapiens may not be able to fix, and even if they could, it probably might be too late by then. There are also possible concerns by the opponents of the technology that plants that have been modified for the purpose of being pest resistant to virus may, or could, possibly have negative and harmful effects on humans. Maybe, the thought could possibly be that these modified crops could lead to serious diseases such as cancer. The argument that probably could be put forth by some is that the technology is too early in its stage to rule out these possibilities. It, therefore, is too great a risk to take.

Genetic engineering in the form of human cloning may also be believed, by its opponents, to be a bad thing. The argument possibly could be that though supporters, including scientists, are highlighting the positives, they are possibly and deliberately concealing and covering up the overall wicked and cruel potential such a technology possesses. The technology can, they may argue, be used to create monsters and tyrants, if fall in the wrong hands. It could also be utilized to breed a kind of hierarchal type and class of human beings. There is also the question of the technology exploiting women and animals, as well as the destruction and murder of lives through the waste it involves in cloning animals and human beings. In addition, there is also the potential psychological aspect to the idea of cloned individuals created through genetic engineering. Some may argue that the individuals so created, would be under psychological duress to be like its progenitor (the person being cloned). There may also be concerns about individuality and independence. The possibility that parents and the society, as a whole, would regard these children as manufactured beings with less respect and so, such children may be treated as vehicles to reach unachieved goals of family members and the weird and ambitious goals of some scientists. They may not be able to develop their own independent path.

There are yet other arguments opponents against genetic engineering may demonstrate. There are potential problems associated with this technology. Scientists have not reached a consensus on whether it is possible to create healthy human clones or even create a human clone, at all. Then, also, there may be concerns that even if an individual is cloned from say an intellectual, there is no guarantee that this clone will be, so to speak, the other person in traits and abilities. This is because there are other factors other than, and in addition to, the inherited genes that will influence the expression of the inherent traits in the clone. As pointed out elsewhere in this book, this again could lead to great psychological and mental instability on the part of the cloned child. In the case of his or her parents, they too could be overwhelmed with disappointments in not having acquired their dream(s). As a result of such possibilities, some will argue against the biotechnology.

Finally, opponents are, or may be, wary of this technology for other reasons than the foregoing. There is concern that this kind of technology may also create a new challenge in terms of legislations and family relations. The genetically engineered child or family is a possible threat to the conventional family and could prove frustrating to current laws on the books of given nations. How would the relations among family members be clearly identified or defined, are some of the big concerns that its opponents my proffer? What is there in place to prevent mother child sexual attraction to each other as a result of say, a mom cloning someone she had sexual affinity toward? How would one define incest in such settings, and so on?

I have attempted to shed some light on genetic engineering. Several possible reasons that may be proffered for its support and application have been discussed. In the same light, several points that may be brought to the fore against the technology have also been presented. It is now up to you, my readers, to take your position. I would humbly advise that you continue to read as much as possible as you can on the subject. Further, I implore you to continue the discussions, with friends, read science journals and attend fora (forums) organized to this end, with a view to have a broader and clearer comprehension of the implications of this new technology. It's your future and that of your children that will surely be impacted with this new innovation.

TEST YOUR LEARNING

---

(1) Using information from the essay, additional resources such as your local and/or school library, (a) define genetic engineering. (b) What are the differences between the two main types of genetic engineering, mentioned in the essay?

(2) Would you say there are differences between the author's style of presenting the first essay of the readings compared with the second essay of the book? Explain.

(3) Using information from this essay, additional resources such as your local and/or school library or by interviewing an expert, (a) explain the major concerns you have about genetic engineering? (b) From what you have gleaned, do the concerns about genetic engineering implementation far outweigh its positive impact(s)? Explain.

(4) At the outset of the reading, the author states that he is merely presenting both sides of the issue of genetic engineering as proponents and opponents of the technology might have. (a) Do you agree that the author is merely presenting both sides of the issue? (b) Does he

present any biases? (c) Are there additional arguments that he could have put forward for both sides? If so, what are those arguments? (d) Are there arguments that are put forward on both sides that are irrelevant and do not provide supporting detail? Explain.

(5)) Finally, do you believe scientists should be allowed, and given public funding, to do research to develop and finetune genetic engineering? Elaborate.

ESSAY THREE

IS AFFIRMATIVE ACTION A JUST AND SOCIAL EQUALIZER?

This discussion will be initiated with a couple of questions which serve to point to the theme of this essay and to establish my objective path clearly. This is with the view to elucidate, most prominently, the question at hand and to elicit some serious reflection on the issue as my readers share in this very important matter. This is an issue that has serious implications on a significant number of our "visible minorities" and women. Here are the questions. What are the objectives of Affirmative Action? Are the reasons for which Affirmative Action was introduced, now obsolete or eliminated? It will be demonstrated, in this essay, that Affirmative Action serves an important end. It will, in short, show that Affirmative Action is a just and social equalizer.

Affirmative Action may be defined as a legislative instrument enacted to foster a state of equilibrium in the workforce among different social groups, in the professions, in education and so on. Its objective is to augment the numbers of the traditionally exploited and discriminated against, such as women and visible minorities in North America. In so doing, it ensures that the relative frequency of these traditionally discriminated against groups comes more on par with the "conventional dominators" of the more important aspects of the workforce and management structure of the North American and European economies.

Having stated the foregoing, it becomes necessary at this point to define the phrase "just and social equalizer." The term "just" connotes fair or upright. "Social" as it is used here, signifies relations among classes of different individuals. We could, therefore, safely label the term "a just and social equalizer" as that which promotes fair relations among different classes of people.

We now move to the reason and need for the implementation of Affirmative Action. It is common knowledge that quite a significant number of minority groups have been isolated not merely from the means of production in the past but also from certain kinds of the so-called white-collar jobs and professions. Some of whom are Jews and blacks, among others. For example, it was not until the early to mid twentieth century that Jews were even given the right to own land in Canada. This kind of unjust affair was not peculiar to that group but equally pervaded other ethnic and minority groups. With the passing of time, coupled with the conscience of contemporary man, some kind of strategy was evolved to right this wrong. It was accordingly that Affirmative Action was born. There was an obvious disadvantage in representation in certain vital areas of the economy and professions for certain groups. Consequently, it was considered, and rightly so, that the time for change had come. The traditionally deprived and discriminated against would get an opportunity to be represented in the professions and play important roles in some of the more recognized segments of these societies. They finally were to be compensated for the wrongs meted out to them for many centuries!

In the current situation which obtains, the question may be asked, have the causes that led to Affirmative Action, completely eradicated the challenges minority groups and women face prior to its establishment, nullifying the injustice which prevailed before its existence? I suspect not. It was frankly, but shockingly, revealed by Canadian statisticians that in the year two thousand one (2001), over forty six percent of the Ontario Province of Canada comprised individuals of ethnic origins. But a mere glance in our Ontario professions and institutions of eminence does not reflect a proportional, parallel growth of these minority groups. The number of black faces in our professions of status continues to lull behind. A similar state of affairs obtains in our universities and higher places of learning. The argument by some critics, however, may be that since Affirmative Action does not solve the problem of inequity and discrimination, it should be destroyed. The counter to such an argument would be as follows. The fact that the "watchdog" system of Affirmative Action is not one hundred percent successful does not make it a failure. A pupil who earns a cumulative average of seventy percent in his or her studies cannot correctly be classified as failing. On the contrary, it becomes even more vital that such a pupil should remain in school and be encouraged to attain to his or her maximum capacity. So it is with Affirmative Action. It probably is not one hundred percent effective in getting the desired amount of traditionally deprived and discriminated against, into the professions, but it does improve their prospects for so doing.

It would seem that it is more crucial and imperative, at this time, that Affirmative Action be allowed to remain and be the watchdog for the so-called underclass and discriminated against in these societies. In light of the reality that a system such as Affirmative Action, is currently in place and the powers-that-be, along with the elitist Caucasian forces, still find the means and ways to discriminate against these groups as is reflected in the hiring practices in our professions of stature, the pertinent question is, what would be the result if the governments of these nations had given the powers-that-be, free rein to openly discriminate and victimize our women and visible minority groups, with the collapsing of Affirmative Action? The possible answer to this question makes even my stomach churns in gruelling pain!

The reality is that our societies, despite our idealist and well-intentioned goals, are still rife with racism and distrust of women and individuals of colour and worse, the subtle and covert view that people of colour are inferior. The black North American experience, in particular, is not very different from the situation that obtained in the Martin Luther King era down South and elsewhere in America even though superficially, this argument may seem nonsensical to many who read it today or may even deny this reality. The respective governments of each region have put legislations in place that make the poor and discriminatory treatment of coloured and women illegal. But what is the real experience of most negros and women today? Who are the housemen in our hotels today? Who are the hotel room maids? Compare this situation to the front desk workers in most of these hotels and so on. In our North American schools, in places like Toronto, for example, even in the so-called black communities, what percentage of the teachers is black? Why is it that this author, having trained as a science teacher and having taught for over eleven years in the Caribbean, served as science department head for three years prior to migrating to Canada, is yet to be employed permanently in a teaching position in Toronto? Why is it that after being licensed to teach in Ontario since February 19, 2003 and having been successful in their required interview every single year to be placed on their so-called Eligible-To-Hire List (foreign trained teachers must be on this list to be hired permanently) and having worked two successive years in so-called Long Term Occasional teaching positions, and having been praised and commended by vice principals and principals for a great job done, I am yet to acquire a contract position? Further, the question must be asked, "Who dominates the administrative levels of these schools?" Why is this so, even in Toronto where more than forty six percent of its provincial population is of ethnic extractions, as mentioned above?

In addition to the above challenges, the events that have taken place on September 11, 2001 at the World's Trade Centre, the well known 9/11, the so-called terrorist activities in Britain, and elsewhere, subsequently, have not helped the cause of minority groups in recent times either. It, frankly, has frustrated and steepened the challenges of these traditionally discriminated against groups. In a way, it seems that at this time, policies, and watchdog institutions such as Affirmative Action become even more relevant and needed to uphold fair practices in employment and university admission practices in these parts of the world.

Judy Wubnig, in her article, The Merit Criterion of Employment, suggests that Affirmative Action, which incorporates among other factors, the notion of compensating present members of a certain group discriminated against in the past, as pointed out above, is wrong. She argues that it is possible to compensate a member of a group only if that member was the one discriminated against. It is my expressed view that the author's argument is, at best, shortsighted and at worst, illogical. Wubnig needs to be reminded that individuals do not exist as a vacuum. Each individual is a microcosm within a macrocosm of individuals. If one's grandparents were deprived of the opportunity to socio-economic growth, this in turn could result in a number of adverse factors for that group or family. The chain effect could inevitably be as follows. Firstly, the father of the grandson would very likely experience socio-economic handicaps as there would have been no base from which he could launch himself. The consequence is severe suffering and impoverishment on the part of this father. The fate of the grandson is very likely to continue along the path of the father for the same reason alluded to. It therefore means that compensating the grandson through Affirmative Action would serve as a just and social equalizer. It is a just equalizer not merely because he now gets an opportunity to participate in processes otherwise denied to him and his kind but by being given this opportunity is an indirect reimbursement to the grandparents and indeed, to himself.

There are other ways that Affirmative Action serves as a just and social equalizer. One is discussed by Pamela Courtenay Hall. In *From Justified Discrimination to Responsive Hiring* (PP 239-241), the writer makes reference to the role-model base that is created through the enforcement of Affirmative Action. She cites in her case, the employment of more female professors of Philosophy who could serve as models for observational learning. Although this argument is in respect to Affirmative Action causing more female university teachers to become available as models, this principle could be applied in a general sense. The creation of "decent" jobs for the otherwise socially deprived individuals serves as a motivational tool for other members of this group. Affirmative Action plays a dual role. Firstly, it augments the self-concepts of such individuals. Secondly, it motivates them to attain excellence. This arises from the fact that there are concrete role models to prove that "their kinds" have no lesser abilities than the traditionally recognized elite.

The aforementioned facts demonstrate how Affirmative Action serves as a just and social equalizer in more ways than one. It ensures that the chasm between the discriminated against and the elite narrows. It fosters psychological wellbeing on the part of the deprived. In short, Affirmative Action assists significantly in levelling the playing field. It attempts to right the wrong, or at least it tries to compensate in some way, for the deprivation and injustices meted out to particular groups of people. It promotes self-worth for groups of people who otherwise would feel alienated from a history pregnant with discrimination and deprivation. Finally, Affirmative Action contributes significantly to a just and social equilibrium of relations in respect to desired professions and status among all classes of people.

TEST YOUR LEARNING

---

(1) Using information from this essay, additional resources such as your local and/or school library, (a) define the term, Affirmative Action. According to the author, (b) what are the goals of Affirmative Action? (c) Are these goals reasonable? Explain.

(2) Would you agree to, or support the arguments furnished by the author, for the need for Affirmative Action in the societies in which it currently exists? Explain.

(3) Using information from this essay, additional resources such as your local and/or school library or by interviewing an expert, (a) did the author adequately and convincingly present his arguments in support of Affirmative Action? (b) Did you identify any biases? (c) Could the premise of his argument stand up to critical scrutiny? Explain in each case.

(4) On a scale of 1 to 10, 10 being the best, 1 the worst, (a) how would you rate this essay?

(b) How would you rate his introduction to the reading using the same scale? Explain. (c) Are there any supporting details that the author provides, that were not relevant to the discussion or in making his point? (d) Could he have provided more supporting details to strengthen his arguments? If yes, what arguments could he have included?

(5) If you are not, or were not a woman, or visible minority, do you think you would have the same views on the question of Affirmative Action that you have now? Explain.

ESSAY FOUR

WAS MARTIN LUTHER KING JR. A TRUE HERO?

In this essay, I will attempt to shed, or recast, some light on the great Dr. Martin Luther King, Jr. In so doing, I will give a brief insight into the background of this great leader, especially for our younger readers. This will include information on where and when he was born, who his parents were, some of the struggles he encountered in the Civil Rights Revolution (Movement) for black Americans, some of his victories as well as some of his defeats and where, when and how he died. I will also attempt to answer the question, "Was Martin Luther King Jr. a true Hero?" It is hoped you will be intrigued.

Before launching into the background and experience of Dr. Martin Luther King, Jr. as promised, and to answer the question posed in this essay, it is important to define and establish what we mean by the term "a true hero". The term "hero" according to the Oxford Dictionary of Current English (New Edition), defines the word hero as: A person noted or admired for nobility, courage, and outstanding achievements. The word "true" as used in this context and also as expressed in the same referenced dictionary, signifies "genuine". In the same breath, the dictionary being quoted describes being "noble" as being "of excellent character". Put another way, the question may also be posed as, "Was Dr. King, Junior perceived by all blacks as a genuine person of nobility, courage and outstanding achievement?" In the section in which I will attempt to answer this question, "Was Martin Luther King Jr. a true hero?" I will endeavour to test whether Dr. King Jr. had actually and satisfactorily passed each stage of nobility, of courage, of outstanding achievement, and character through our litmus as will be presented in this paper. However, as explained in the foregoing, I will, at this point, outline some background on this famous and world acknowledged and recognized stalwart, for my readers who may not be as cognizant of the history and deeds of Dr. Martin Luther King Jr.

WHERE AND WHEN WAS MARTIN LUTHER KING, JR. BORN?

Dr. Martin Luther King Jr. was born on January 15, 1929 in Atlanta, Georgia, U.S.A. of parents of middle class means. It is interesting to note that Martin Luther king Junior's name was

Michael Luther King Jr. at birth and up and until he was about four years old, this was the case. For some family reason, having to do with a mix-up of names with another close family member, his dad decided to change both his and his son's name from Michael to Martin to avoid any confusion with their relatives who bore the same name. Some say that King Senior did the change of name from Michael to Martin, out of respect for his elder relative of the same name. Even though at the time of his birth Dr. King Junior's parents were middle class blacks doing much better than most blacks and some whites, they were not always wealthy. The fact is that Martin Luther King Senior did not acquire much schooling during his early years and was an offspring of dire poverty, like most poor blacks of his time. He was born to a very pauperized, farming family in Georgia in a rural township called Stockbridge. He (the senior King) left his home to get a job to better himself whilst studying at night schools to gain an education. He later got accepted into college in Atlanta, acquired a bachelor's degree in Divinity and became a Baptist pastor (minister of religion).

Martin Luther King Junior's mother was Alberta, and the daughter of a pastor. She was often described as a fine housewife. Most women of means, as she was later to become, stayed home then and took care of the children. Martin Luther King Sr. a minister of religion

(Baptist) at the time of Dr. King Junior's birth married King Junior's mother, Alberta Christine Williams in 1926. The kings later had two younger siblings for King Jr., Christine and Alfred.

MARTIN LUTHER KING JUNIOR'S EARLY EXPERIENCES

Dr Martin Luther King Jr. had a great life economically as a youngster and adult. He had not personally experienced economic hardships but later saw all the poverty around him and the exploitation of blacks and the deprivation and degradation of their rights as human beings. His parents also instilled and encouraged a sense of responsibility in Martin Luther King Jr. since his childhood days. He earned materials such as toys by working and accumulating money to pay for them. He, however, despite his affluence, experienced racism and scorn which was prevalent in the South then due to special laws, the so-called Jim Crow Laws of Alabama. These laws, established after slavery, were enacted to ensure that blacks would remain the underclass and were unable to mix freely with whites who were considered the superior race. For example, there were separate washrooms for whites as opposed to blacks. The same obtained in restaurants, hotels, and so on. There were segregations of the school system. There were the black schools which offered inferior quality education in most cases, compared with the relatively better, quality schools for whites only. Many blacks were also denied the right to vote. In buses, the first ten rows were reserved for whites only. Blacks could only travel in the eleventh row and beyond. Even then, if the buses were to be crowded, the bus drivers had the legal right to ask blacks to get up for white passengers to sit. It was exactly what occurred on Thursday, December 1, 1955 while one Rosa Parks sat in the section allotted to blacks in a bus in Montgomery on her way from work. She worked as a seamstress. Even though she was seated in the section assigned to blacks, she was asked to get up for oncoming white passengers to sit. Whilst three other black passengers sitting beside her got up, she refused and was subsequently arrested. It was this cause, started by Rosa Parks' quiet protest that Martin Luther King Jr. was to later triumph and got the Civil Rights Movement in striking motion, beginning in Montgomery with the desegregation of the bus system. It is interesting to note that the first set of black slaves landed in the USA in 1619. This was to be the start of the challenging onset of the black people of the USA and North America in general. A struggle which still continues, to this day as mentioned in the foregoing essay in this book.

Martin Luther King Jr. is what we would probably define today as a "gifted" child. He attended school as early as five years old which was a year and one-half earlier than the established school going age at the time. However, this was later discovered, and his schooling was terminated by the authorities. Martin later was re-admitted at the correct age and did well academically, though somewhere at the secondary school level, he started to discover that what he was being taught in school might not have been quality education and probably was information that at his age, he should have been taught a few years earlier, compared to his white counterparts. Despite the relatively lower standards in black schools compared to white schools, Martin Luther Jr. attended the top black schools that were available at the time. Some of the schools he attended were Yonge Street Elementary School, the Atlanta University Laboratory School, Booker T. Washington High School, Morehouse College and Boston University. At Morehouse College, he acquired a bachelor's degree in Sociology. It is important to note that Martin Luther King Jr. passed the entrance examination to this college whilst he was still a high school junior, despite the disadvantage in the black school systems.

In addition to the above, due to the inferior lessons he had at high school, he initially had problems in college as he was merely reading at a grade 8 level and was unable to write sentences. He doubled his efforts to learn how to read much better and developed a great love for all types of reading materials. He also developed a fondness for big words and "word organizations and structuring" and soon became a class act in oratory and written presentations! Whilst he studied theology later, he would also find time to read Law, Politics, Philosophy, History, and others. He developed a liking for the work and vision of Mohandas Ghandhi of India and David Henry Thoreau. These individuals stipulated a theory and doctrine of nonviolent interactions with one's enemies to overcome and conquer them. As a matter of fact, this was the strategy that Mohandas Ghandhi employed to get independence from Great Britain in 1947. It should also be noted that immediately after Martin Luther King Jr. graduated from Morehouse College, he got accepted to Crozer Seminary where he studied theology (Bachelor of Divinity degree) and later acquired his doctorate at Boston University at a relatively young age, 26 years old. It must also be noted that at the Crozer Seminary, it was the first time that he would attend school with both blacks and whites. Whites comprised sixty six percent of the school's enrollments, the rest being black. Most, if not all, of Martin Luther's struggles for Civil Rights for American blacks, in particular Southern blacks, were ones that advocated nonviolent strategic approaches such as sit-ins by blacks in white only restaurants, freedom riders, etc., until they were served and/or recognized. It is also interesting to note that when Martin Luther King Jr. entered Crozer Theological Seminary, Chester, Pennsylvania in 1948, he met two professors Dr. A. J. Muste and Dr. Mordecai W. Johnson whose teachings about Gandhi were later to influence his style for the Civil Rights struggle as mentioned earlier. These two scholars taught King Jr. of Gandhi's teachings of nonviolence necessary to overturn the injustices of the world. He was later to travel to India to get firsthand information on the nonviolent approach.

SOME STRUGGLES

Martin Luther King Jr., although he had enjoyed great successes, did have some struggles as well. Some black leaders were very critical of his nonviolent approach to achieve equality and equal rights for American blacks. Some of his critics were Malcolm X (formerly known as Malcolm Little), Stokely Carmichael and Robert Williams. Robert Williams was then the local President of the National Association for the Advancement of Coloured People (NAACP) for the Monroe County, California Chapter. He was described as a diehard supporter of militant retaliation on the part of blacks. He was also said to have, as an ex-marine, set camps and gave blacks military training to encourage them to defend themselves from white aggressors. This was, of course, in direct contradiction to Dr.

Martin Luther King's philosophy. Later, in this essay, more will be said on the others mentioned in this section.

Dr. Martin Luther King Jr. was also seen by many of his critics as being a womanizer and had been accused of having sexual relations with many women, even while being married and preaching to his church's congregation. These, however, were all allegations not proven or substantiated.

WHERE AND HOW DID MARTIN LUTHER KING JR. DIE?

Dr. Martin Luther King Jr. died on April 4, 1968 at age 39 years, after being shot in the face by an escaped convict, James Earl Ray. James Earl Ray was also a racist supremist. News spread that Dr. King Jr. and others, including Rev Ralph Abernathy and Rev Jessie Jackson were staying at the Lorraine Motel in Memphis where they were in preparation for an April 8 peaceful protest. He was shot while standing on the balcony, intending to talk to an associate on the lawn outside and in front of the motel.

WAS DR. MARTIN LUTHER KING JUNIOR PERCEIVED BY BLACKS AS NOBLE, COURAGEOUS, OF GOOD CHARACTER AND OUTSTANDING IN ACHIEVEMENT?

Dr. Martin Luther King Jr., in my estimation, was of fine class and a true hero of the twentieth (20th) century and for all times. He had done, more than any man, black or white, had ever done for the world in terms of justice, equality, and the rights of humankind. His vision for the poor and underprivileged of America, most of whom were blacks, has caught on to a chain reaction that has led to a number of positive changes and radical achievements that the world has come to see and enjoy today. The mere idea of having had a black president of the United States of America or a woman who had run to be president of the United States of America is a striking example of the legacy of Martin Luther King Jr. in particular, and Ralph David Abernathy, Jessie Jackson, among others.

In addition to the above facts, there are other glaring pieces of evidence to testify to Dr. King Junior's worthiness as a true hero of all time. Even though Dr. Martin Luther King Jr. could have easily preached and instigated violence for Caucasians through the influence he wielded among the black masses, he refused to do so. Even after he had been the recipient of violent assaults and onslaught by his white enemies, beatings and so on, he stood by his conviction to promote peace without vengeance.

The proof of Dr. Martin Luther King Junior's quality as a leader and heroic stature was also obvious in the way he conducted himself being mistreated by the powers-that-be during his time. He had never used brute force or resistance when he was being jailed even for questionable acts by the authorities. As a matter of fact, it is the unique and quality strategic approaches that had earned him the unfavourable title of "Uncle Tom", a kind of subservient slave, and other "not-too-honourable" names, by a few other black leaders at the time. There were a few black leaders who thought that he was "sucking up" too much to whites and that he was exhibiting behaviours tantamount to a feeling of inferiority to the exploiters of the American black people. Those in opposition to his style of leadership and methodology in getting the attention of the authorities thought his way would never serve to get justice, equality, and the rights that black Americans deserve. For example, when Dr. King Jr. failed to acquire meaningful reform for blacks in places such as Chicago, Illinois, and Memphis, his black critics were quite vocal, some would even say, disrespectful of his leadership style and abilities. Among those black leaders who were antagonists to the ideas and approaches of Dr. King Jr. were Chester Robinson, Stokely Carmichael, Malcolm X and Robert Williams. It is alleged that he was openly criticized, if not ridiculed by these black leaders at times.

Notwithstanding the criticisms by some of his black leader associates, Dr. King Jr. had never been bitter toward them nor their opposing camps. Neither did he endeavour to create factions within the black freedom fighters' cause. He knew very well that such acts would only further frustrate the successes that the others had worked so hard and sacrificed to achieve in the South. This, in his view, would only satisfy selfish ends which was obviously not his goal. This, again, depicts an unquestionable and unambiguous indication of the selfless and heroic talent there was in Dr. Martin Luther King Jr. Despite all the differences with these black leaders, he managed, almost single-handedly, to maintain a working relationship with them all, and often times sought their assistance and input in areas not pertaining to the nonviolent aspect of his approach through useful inputs in other respects.

Stokely Carmichael became the Student Nonviolent Coordinating Committee ((SNCC) Chairman in 1966, an organization which was earlier influenced by Dr. Martin Luther King Jr. This created more problems for King Jr. The new Chairman of the organization thought that what was needed was more militancy, in a sense, violent revolution, to demand equality and justice for blacks. He preached retaliation and the exclusion of whites, some of whom had joined the cause through King's appeal and were working side by side with blacks in the struggles for change. Carmichael thought that Dr. King Jr. was begging the whites for what was rightfully blacks', and which had been stolen from blacks. In spite of all the bickering and opposition to

King Junior's peaceful approach in the struggle for equal rights and justice for black Americans, in the end, Dr King Jr. came out the champion. He had achieved incredulously for America's poor, indeed the for the world's blacks and so-called underclass and all people of colour, in general. It is in this regard, that Martin Luther King Jr. must be judged as to whether or not he had satisfied the definition for "true hero". It can be clearly gleaned from the above, and by reading the various historical accounts, that Dr. Martin Luther King Jr. had satisfied all the criteria for nobility, courage, and outstanding achievement. In short, Dr. Martin Luther King Jr. is one of, if not, the most deserving hero of all time. His life and works should be taught in all the schools through all levels, on this planet, and any future schools or planets other than earth, conceivable in the future!

TEST YOUR LEARNING

(1 (a Using information from this essay, additional resources such as your local and/or school library, explain whether the author's presentation included any bias or biases? (b Were you impressed, or not, of his seemingly novel approach to Martin Luther King Jr.? (c Do

you agree that his approach to the topic was a novel one? Explain in each case.

(2) Would you agree to, or support the arguments furnished by the author that Dr. Martin Luther King Jr. be considered a true hero of all time? Explain.

(3) Using information from the essay, additional resources such as your local and/or school library or by interviewing an expert historian, did the author adequately and convincingly present his arguments in support of Martin Luther King Jr.? (a) Did you identify any biases? (b) Could the premise of his argument stand up to critical scrutiny? Explain in each case.

(4) (a) On a scale of 1 to 10, 10 being the best, 1 the worst, how would you rate this article? (b) How would you rate his introduction to this essay, using the same scale? Explain. (c) Are there any supporting details that the author provides, that were not relevant to the discussion or in making his point? (d) Could he have provided more supporting details to strengthen his arguments? If yes, what arguments or points could he have included? (e) Could he have supplied more opposing points of view? Explain.

(5) If you are not, or were not, a black person, or visible minority, do you think you would have the same conclusions about Martin Luther king Jr. as you have now? Explain.

ESSAY FIVE

ARE JAMAICANS AND BLACKS IN GENERAL REALLY EMANCIPATED?

In this section of the reading, dedicated to my homeland, Jamaica, I raise the question, even as I encourage Jamaica to celebrate its two most salient dates of August 1, 1838 and August 6, 1962, its Independence and Emancipation Days, respectively, which I have juxtaposed and referred to as 'emancipendence', (a phrase originally coined by Jamaican Journalist, Michael Burke) as to whether Jamaica and the working masses, at home and abroad, have actually seen the back of slavery, so to speak. I argue that slavery is still alive, rampant, and kicking, and has merely evolved in a contemporary camouflage, disguising itself as a trick to the many unsuspecting, deprived masses. I touch briefly on how this sad state of affairs has led to other problems of school truancy and criminality on the part of the offspring of these modern, enslaved workers. I speak to the various forms that this contemporary slavery has acquired and that we should be mindful of them with a view to bring about required, positive change. This change must be envisioned and understood to be possible.

To make sense of some of the references in this article, it should also be noted that this article was written on July 25, 2006 and published in part on August 6, 2007 (Jamaica's Independence Day) in the Jamaica Gleaner, just a few weeks prior to Jamaica's Election Day of August 27, 2007. The purpose of this section of the book is basically to facilitate some new and refreshed thinking on the question of the accomplishments, or lack thereof, of not only black Jamaicans but that of all minority groups throughout the length and breadth of the globe, irrespective of where they call home presently.

In Jamaica, August 1 and 6 are celebrated each year as Emancipation and Independence Days, respectively. All reasonable individuals of Jamaican roots would naturally have no qualms with our recognizing these days. These two dates; August 1, 1838 and August 6, 1962 MUST be two of the most salient dates in our nation's history. The first brings, technically, to culmination, a period of horrific spate of exploitations of a subjugated and vilified masses hitherto unconstraint. A system that deprived thousands of homo sapiens (human beings) of their cultures, their religions, their freedoms, their hopes, dreams, and even their names, among other things. The second date serves to establish sovereignty not merely from our so-called motherland, Britain, but importantly, should involve sovereignty of thought, of will and the ability to think and act independently even as we seek some semblance of alliance as we are being weaned, as in the case with our current arrangements with British police in the fight against crime in Jamaica today.

Our festivities, over the upcoming 'emancipendence' period, should be done with vigour and gusto, despite elections fever of August 27, 2007 blooming in the air but only in the context, and with the understanding that these achievements are in 'working progress'. We, as a people, must realize that we are still struggling for true and genuine emancipation of slavery and total independence for our country and people of colour, as a whole in Jamaica and abroad.

In 1948 the Universal Declaration of Human Rights, adopted more than sixty years ago, proclaimed that "no-one shall be held in slavery and servitude" However, even with the drafting and implementation of this universal document, legislated approximately three scores ago, and even though slavery was said to be abolished in Jamaica land we love on August 1, 1838 and our independent nation came on stream later on August 6, 1962, we would be naive to be blindly coaxed in the belief that subjugation of the masses and significant droves of exploitation are non-existent in contemporary Jamaica and abroad, even as we celebrate and recognize our proud accomplishments and prepare to exercise our franchise on August 27, 2007.

The problems of slavery and slave labour continue to circumvent and permeate the lives of ordinary Jamaicans, at home and abroad. For example, the alleged working conditions that permeate the Farm Work Programs in Canada and the United States may be solid testimonies that slavery may yet linger unrelentingly but colourfully dressed in its contemporary disguises and brainwashing trickeries and camouflages. Farm workers, it is alleged, are tricked into enslaving themselves for feeble wages whilst fattening the coffers of the great "plantocracies" and often, though there have been some improvements, the allegation is that they exist in less than hospitable and humane ambiences. Farm workers, for example, upon their being recruited, the allegation goes, are still required to strip naked in some cases in the full, glaring glory of their colleagues, to test for their fitness for this slave-like endeavour that they must endure for six months annually! How different is this, if it is true, from the "backra massa" days?! The situation that obtains with other workers including so-called professionals is not much better. In particular in places like North America and Europe, the working conditions of large droves of workers are nothing short of slavery! One has, not even the time, in these circumstances, to enjoy one's so-called 'acquired properties' due to the urgency of work from 9 to 5, and the need to pay back overburdened and over-taxed loans and mortgages, if one is lucky to 'own' her or his home! In many cases, one runs from one job to the next, reaching home at ungodly hours at nights or at the sunless morning's dawn, on a regular basis. The children in these cases are forced to fend for themselves and to choose their own forms of entertainments. Entertainments that sometimes lead to them playing truant from school, engaging in criminal activities, or simply ending up in jail! These are actions that promote the vicious cycle of poverty, victimization, and enslavement on the part of coloured peoples, here and abroad.

The challenge is that we have been schooled to perceive slavery only in the context of images of the Middle Passage, black skin folks with physical shackles around their ankles being handed a machete and a hut to 'work' for free. To most Jamaicans and many individuals of African descent and extractions, from home and abroad, slavery is simply, but wrongly, explained as an anachronism or attribution from a long and atrocious past. The reality that the majority of blacks, working blacks, and other people of colour, after so many years of the so-called abolition of slavery all over the globe, are still unable to enjoy the fine things of life even though they work so hard, has strangely not elicited this new awakening! This reality has not managed to encourage us to pose the question about this farce about abolition. It is a strange reaction by our peoples. It is a reaction induced and moulded by a careful and deliberate strategy, organized, and implemented via the powers-that-be. It is as a result of a well-orchestrated brainwashing tool so blatantly at work almost everywhere, but in particular, in North America and Europe. However, this myopia in our line of thinking and vision is very detrimental to the true eradication and non-exploitation of a great many folks of colour, locally and abroad.

Slavery has evolved and has now taken on a contemporary trend. Some years ago (2006), an example of the kind of modern form of slavery came to the fore in the allegations of the Jamaican deacon with the teenagers performing coerced, sexual acts for the cameras, driving around in the church's bus. This of course, would appear that its intent was for the pornography market of the Internet and other such media. Then there are the poor, industrial workers who spend most of their time at work as alluded to earlier, and away from home and family but still are unable to meet their basic needs of providing adequate food and shelter as well as basic, affordable education for themselves and children. In the case of immigrants in the USA, Canada, and most parts of Europe, even if he or she has resided in his or her new North American and European home for decades, has committed no crime, he, or she still does not even have the right to vote, unless he acquired citizenship status. He has no say in the direction of the country in which he or she contributes. He or she has no say or control in the kind of education in which his or her children will partake. He or she has no voice. He or she, in essence, is the contemporary slave. He or she has the right to work but must remain voiceless as is demonstrated in the blatant non-right to vote and to take part in the affairs of a country that he or she might have even lived longer than most of its citizens! In such a stead, can blacks and other people of colour truly say that they have been emancipated? Rosa Parks and Martin Luther King Jr. had managed to end segregation in America, but are black Americans and coloured peoples of the globe truly incorporated in the main thrust of things? Can we truly say that there is complete and full integration of the races in these parts of the world?

It is the common belief that some young people, including Jamaicans, get caught up in what are known internationally, as "debt bondages". A person enters in a debt bondage arrangement when her or his labour is demanded as a means of repayment of a loan or money given in advance. These situations usually arise as people, in general, mostly feel that the grass is always greener on the other side of the opaque wall. So, they gamble by agreeing to proposals, most of which are illegal and "con-artist" in intent. Usually, these individuals are brainwashed, as in the human trafficking scams, into working into various kinds of jobs, mostly illegally, abroad. The person is either given or promised a portion of money or valuables. These individuals are usually put in a situation that it is impossible for them to pay back these loans and are usually drugged unknowingly. They are, in essence, enslaved!

In light of the foregoing, as we celebrate this 'emancipendence' season upon its arrival, let's not go overboard with it, but use this period as a reminder that we have come a long way. Let's employ this achievement as fertilizers to induce and elicit a more equitable relationship among the haves and the haves-not. Let our motives in celebration be for true 'emancipendence' of all the exploited, on planet Earth, and the complete eradication of slavery here and abroad. Let us, once and for all, be in the position to scream, from the bottom of our hearts, "Happy 'Emancipendence!'" but this time with a clear conscience, knowing this to be factual and not some usurped notion!

TEST YOUR LEARNING

---

(1)Using information from this essay, additional resources such as your local and/or school library, (a) explain whether the author's presentation included any bias or biases? (b) Would you agree that the author's approach to the perception of slavery in Jamaica and elsewhere, is a novel one? (c) Were you impressed, or not, of his seemingly novel approach to the question of slavery and emancipation? Explain in each case.

(2) Would you agree to, or support the arguments furnished by the author that black Jamaicans and blacks and other minorities in North America and Europe as a whole, are still being enslaved? Explain.

(3) Using information from this essay, additional resources such as your local and/or school library or by interviewing an expert historian or journalist, (a) did the author adequately and convincingly present his arguments in support

of his views? (b) Did you identify any biases? (c) Could the premise of his argument stand up to critical scrutiny? Explain in each case.

(4) (a) On a scale of 1 to 10, 10 being the best, 1 the worst, how would you rate this article? How would you rate his introduction to the article, using the same scale? Explain. (b) Are there any supporting details that the author provides, that were not relevant to the discussion or in making his point? (c)) Could he have provided more supporting details to strengthen his arguments? If yes, what arguments or points could he have included? (d)Could he have supplied more opposing points of view?

(5) If you are not, or were not, a black person, or visible minority, do you think you would have the same conclusions about the existence of slavery (or the article), current or in the past, as you have now? Explain.

ESSAY SIX

IS COMMITTING ABORTION, COMMITTING MURDER?

The question of abortion has been discussed at many fora (forums) and dinner functions and other such gatherings. As a matter of fact, where there have been frequent interactions with two or more people of varying economic status, education, morals, values, religious affiliations, and parents from different walks of life, etc., the topic is usually a hot one. Parents, children, and siblings have had heated debates/disagreements on the question on whether or not one should be free to commit abortion, if one feels like so doing. Religious and political institutions have had their fair share of debates and discourse on the subject as well.

I will attempt to present both sides of the arguments, both for, and against, abortion that might or might not have been hurled around for many decades. You will decide if committing abortion is in fact murder. To give a deeper sense and comprehension of the probable arguments that may be put forward for, or against, abortion, and to establish and fortify my readers with the requisite knowledge to arrive at an informed position on the issue, I will define the terms abortion, murder, embryo, foetus, what it is meant to be "moral" or "immoral" and some science based and "non-science" based terminologies as well as identify the two main types of opposing participants concerned in this debate. I will also shed some light on the bases of the philosophy of these two opposing views on the subject of abortion so as to assist my readers in informing their opinions on whether it is moral or not, or whether it is simply, the committing of murder on the unborn child by those who support and carry out such acts. My intention, in this paper as in some in the previous pages of this book, is simply and merely, to provide information from both sides of the debating divide on these issues, and to allow my readers to come to their own, informed conclusion, thereafter.

Before launching on some basic definitions such as what is murder or abortion and so on, let's examine, define, and identify the groups that are either in support of abortion or are against it. In so doing, also, let us examine the original premise on which their points of view are set. In addition, and very importantly, let us also endeavour to understand their reasoning from different perspectives and be open-minded in so doing.

The debates surrounding abortion, whether in support for the act or in the furnishing of arguments against it, are usually emanated from so-called advocacy organizations. These advocacy bodies usually comprise two unique but opposing groups of individuals who offer strong arguments to support or demonize abortion. In most places, in particular the USA and Canada, those who perceive themselves as opposed to abortion fall in a group that are coined, pro-life and those against the legal restrictions on abortion are said to be pro-choice.

WHAT DOES IT MEAN TO BE PRO-LIFE?

Pro-life is a terminology employed in various circumstances and situations. It may be used in relation to circumstances that are linked to human embryonic stem cells, euthanasia, human cloning, and abortion, among others. It is a phrase that symbolizes many perspectives and activist movements in bioethics. Put another way, it is a term that is utilized to oppose practices such as human cloning, abortion, euthanasia, and so on. In short, it speaks to the political and ethical viewpoints that all homo sapiens (human beings) have the right to life and that this right to life incorporates that of the embryo and the foetus as well.

WHAT DOES IT MEAN TO BE PRO-CHOICE?

Pro-choice is a terminology utilized, in most cases, though sometimes to others, specifically in relation to a woman's pregnancy and her right to terminate, or not, such a pregnancy without being restricted legally or otherwise. It is chiefly used in relation to a woman's right to abort a pregnancy, or not, as she sees fit, without having to endure legal ramifications or challenges. In short, it stipulates the ethical and political viewpoint that a female should have the unrestricted right and full, free, and complete control over her pregnancy or fertility. It should be noted that this right, incorporates and comprises the idea that the protection and security of reproductive rights are naturally a woman's. Reproductive rights include availability and access to factors such as sex education, safe and legal abortion, fertility treatment, protection against coerced abortion, contraceptives and so on.

PRO-LIFE BASIS

Pro-life advocates argue from the point of view that they have to protect the unborn and individuals who because of incapacity cannot make the choice on whether they should live or die on their own. This protection which incorporates the embryo, foetus, and human beings who are so weak or incapacitated that they are incapable of preventing acts such as euthanasia (assisted suicide) being effected to them. One segment of this group, in particular those from a religious base, supports the concept of pro-life from a religious perspective, especially from the Christian bible's teachings.

PRO-CHOICE BASIS

Pro-choice advocates argue from the point of view that it is their bodies and they have the right to do as they please. No one, including the government, should stipulate to them whether they should bring a child into the world or not.

This is to say, that it is the pregnant woman who will have to experience all the joys or sorrows that are associated with the choice to abort or not. In a sense, they are stipulating that what they do is their business and nobody else's. This point of view may or may not have to do with a non-religious viewpoint or foundation or a non-recognition of religious teachings.

Having outlined the above, I will define some of the terms that you may find useful in assisting you in determining whether your position is one to support abortion or not to support the act. Here are the terms and definitions.

ABORTION

Abortion is the elimination or expulsion of the embryo or foetus from the uterus or womb of a woman, resulting in or caused by its death. Put another way, abortion is the process in which the developing child is removed from a mother's womb for the purpose of terminating the pregnancy prematurely and getting rid of the potential baby.

EMBRYO

The embryo is what is formed when a male gamete (sperm) fuses with a female egg (ovum). This can be brought about by natural means, for example, during copulation (sexual intercourse) in animals including homo sapiens or by artificial means such as In Vitro Fertilization (IVF) where the sperm of the male and the egg of the female are collected and placed in a test tube or Petri Dish and manipulated to bring about fusion (fertilization), then implanted in the uterus (womb) of the woman to bear the child. Note, this "mother" can be the mother who contributes the egg or a surrogate mother who may or may not be related to the child of which she will give birth.

IT MAY ALSO BE USEFUL TO REVIEW THE FOLLOWING FACTS WITH RESPECT TO THE EMBRYO ALREADY MENTIONED IN A PREVIOUS SECTION OF THIS BOOK:

All human beings (homo sapiens) and other animals, as discussed in the foregoing, are made up of somatic cells. It is also a scientifically established fact that all somatic cells are related due to their common origin of the single cell that evolved over a billion years ago. During fertilization (the fusing of the sperm with the egg) a single cell is generated to form that which is known as the zygote. The zygote rapidly divides during the first week of fusion and onwards (fertilization) to form a ball of cells, not discernible by the naked eye, known as the blastocyst. It is sometimes also referred to as the pre-implantation embryo, a fact that speaks to the reality that it is not yet implanted in the uterus (womb) at this stage. It is interesting to note as well that scientists have discovered that approximately forty percent (40%) of all pre-implantation embryos developed at the stage of normal sexual reproduction does not get embedded in the wall of the uterus (womb) and is therefore destroyed as a consequence. It is also important to note that scientists have also established that at the blastocyst stage, there are no somatic (body) cells in existence or formed and there is no indication at this point either, for any remote signs of differentiation of the cells of this small, microscopic ball roaming around in the womb (uterus).

Finally, the following scientific facts should be noted by readers. The embryo is not implanted in the uterus (womb) until about fourteen (14) days after fertilization (the fusing of sperm with egg). Beginning at the time of implantation to the wall of the womb at, and subsequently to the fourteen-day period, the embryo begins to form what is described by scientists as primitive streaks. The primitive streaks formed can be one, two, three or more. These primitive streaks are a scientific indication that the embryo cells will form an individual homo sapien (human being). If there is only one, single primitive streak formed, the embryo will develop in a single, individual adult. If the embryo develops two primitive streaks, it grows into identical twins. Sometimes, and very rarely, the embryo develops two separate primitive streaks, but these streaks do not become completely separated. In such an eventuality, conjoined twins (Siamese twins) are formed. Even more amazing and even more rarely, there are cases in which two separately fertilized ova (eggs), adhere themselves together to develop a single embryo with two different types of cells and sex types. This results in a single homo sapien (human being) with some of the cells in their body being female from the original female embryo and some cells being male from the original male embryo. This is described scientifically as tetragammetic chimera

FOETUS

The foetus is the stage of development of the pregnancy that comes after the embryo stage but before the birth of the baby. In homo sapiens (human beings), the foetal stage of development occurs or starts at the end of the eight-week period after fertilization. It is at this stage that the major structures and organs are formed or begin to be formed.

MURDER

Murder is the illicit destruction of life by one, or a group of persons over another which may be a single individual, two, or more persons. It is interesting to note that murder is different and separate from all other types of homicides in that murder entails the elements of intent and a deficiency of justification.

MORALITY

Morality refers to the notion of human action which pertains to matters of right and wrong or good and evil. The matter of what is right or wrong is a subjective one and is mostly determined and decided upon by the group, community, or the structure of which one is brought up and of which the member forms a crucial component. Morality can be distinguished into different facets, such as individual distinction, system of valued principles and so on. (Kindly do research in your local libraries to inform yourself more in the areas of morality). There is what is also known as group morality which is morality that emanates out of a feeling of duty and obedience to the goals, rules, and norms of the group to which one is a part. These norms and rules may or may not be stated but are understood to exist and are the requirements to be satisfied for consideration as a part of this structure. In a sense, members are codified to behave in accordance to the concepts, beliefs and culture of the given group. Accordingly, one's morality is unique and peculiar to one's culture or structure of which he or she is a component. Put another way, what is right or wrong for one culture, tribe or institution may not be similarly perceived by other cultures, tribes, or institutions. In addition, one may act in a way that subscribes to his or her component culture, tribe, or institution, even though he or she may not have any conviction or genuine belief in its moral basis.

SOME ARGUMENTS THAT PROPONENTS OF ABORTION MAY OFFER.

There are several probable arguments that proponents of the right to abortion may present and might have, in fact, presented in the past. The first argument is that as a woman, no politician, legislature or anyone for that matter should have the right to dictate to a woman as to whether she should or should not abort a pregnancy, irrespective of her reason for wanting to do so or not. Her point is that it should be nobody's business. She may argue that she will have the sole duty to live with any decision or not that she considers in respect to such a situation and no one else. Secondly, she may argue that if she does not fight to uphold this right there is no guarantee that the powers-that-be will protect her or her child in exceptional cases, for example, where she may become pregnant due to being raped or due to ill-health, is incapable of bringing a healthy child or in situations where it is determined that to carry a child into this world may result in critical and lasting suffering for that child. What if this child (foetus) should develop HIV/AIDS, cancer or should have some type of disease that is medically diagnosed to lead to physical and psychological pain, how would she guarantee the right to abort such a pregnancy, may be some of her concerns? She may also want to have that right to abort in situations where she had been tricked and got into a relationship with a person who has already had family commitments of his own but had gotten her pregnant without her having cognizance of his other relationship. To avoid problems for her child, herself, and the other family, it may be in her interest, in such situations, to abort the pregnancy, she may argue. She needs to ensure, through having the right to abort, that that right of pro-choice is always available to her, if, and when, it is needed.

There are other reasons that may be made, and that have probably been made in the past, in support of the right to abortion. The reasoning may be presented that if the full and free right to abort is curtailed or tailored in any fashion, this could lead to specific problems for women. Say, for example, the law was limited to abortion to be done only under certain conditions, who would decide what those conditions are and even after those conditions are identified, will a uniform and standard yardstick be used to measure and determine each, unique situation?

In addition to the above arguments furnished, proponents of the right to abort their babies, or not, are pro-choice in outlook. This means that the woman, or any person for that matter purporting this view, believes in the right to choose abortion or not without hindrance. Being pro-choice, by philosophy, means that these women (persons) construct their belief systems in the perspectives of individual liberty, reproductive freedom, and reproductive rights. In essence, what this means, is, that women have achieved a level of contemporary physical and psychological advancement and justice, separating them from the old days of subservience, may be the strong point. These are in contrasts to that which obtained under the old plantocracy system where women, minority groups, and blacks were subject to the whims and fancies of the authority of the day, most of which were imperialist in nature. To give up this right, by those pro-choice vanguards, is tantamount to turning back the hands of time and accordingly, in their view, is a backward, retrograde step.

There are still other arguments to support the pro-choice philosophy. The right to abort or not to abort has certain inherent elements intertwined in it. Some of these bundled rights would include the right for the woman to be educated in terms of education in sexual matters and so on, so she can make an informed decision on matters related to the issue of abortion. If this right is forfeited so would be these inherent, bundled rights. The same would be true for programs that have to do with contraception availability to those who cannot afford them and so on.

SOME ARGUMENTS THAT OPPONENTS OF ABORTION MAY OFFER

There are various and numerous arguments that opponents of abortion may put forth and have probably put forward in the past. The first is that the life of homo sapiens (human beings) which, according to their teachings, includes the embryo and foetus, must be valued, and protected until natural or normal death occurs. This viewpoint is embedded greatly in a religious, mostly Christian, theological foundation. Based on this Christian teaching, it is immoral to commit abortion and worse, it is committing murder to so engage. Secondly, the opponents of abortion will argue that nothing can justify the taking of a human life. It does not, and will not, explain away abortion for special purposes such as a baby produced through the mother being raped. Neither will they be sympathetic to arguments that a potential child would be useless as a person, due to disease infestation and so on. In this case, the opponent to abortion will argue that the child is God's will and should be allowed to enjoy life. It will also be pointed out that God has control and power over everyone and everything and accordingly, even a critically ill foetus, having the potential to be a very sick child, should not be murdered, would be their discourse. Put another way, any intentional destruction of a human life irrespective of its stage of development or status in life, is immoral and unethical, worst sinful, according to this line of reasoning.

There are other points that opponents to abortion would proffer. Some opponents to abortion may and probably have argued in the past, that there is no need for a woman to become pregnant if she does not want to have a baby. The argument would be that there are numerous methods to prevent this from occurring. The woman can apply abstinence. In this situation, there is no possibility that she would become pregnant with an unwanted baby. Those who do not have the will to abstain from sexual activities could embark upon any of several forms of contraceptives on the market. Even in the cases of rape, there is the day-after pill which ensures that the woman does not facilitate a fertilized, implanted embryo, the stage at which life begins, some may argue.

In addition to the forgoing points, there are others that could be made against abortion.

The opponents to the act of abortion could also make the point that even if one is not religious, from a medical/biological perspective, abortion should be frowned upon. The point may, or could, be made that there is no biological difference between an embryo or foetus and a fully grown human being. Therefore, to abort it, is to kill a human being which must be considered as murder. In light of this fact, it could also be argued by the opponent to abortion that the aborted child is subject to pain in effecting this act. In such a case, the argument would be, abortion would be immoral and unethical. A third point that may be proffered here is that since there has not been a universal consensus within the scientific community as to when life begins, it would be prudent to embark on the path of caution and not trod on a slippery slope of aborting embryos which may or may not be a person.

Some final points that may be put forward against abortion could be as follows. Abortion could also lead not only to pain and injury to the unborn child as discussed above, but to injury to the mother herself. In some cases, these side-effects may be great enough to take the life of the mother herself or cause her to be so incapacitated that she could also become a burden to the society and even herself. Some may also argue that abortion is tantamount to an act of violence toward the unborn child. The act of abortion and the associated psychological stress, even when denied by persons who have done it, are subject to the manifestation of psychological challenges for the mother and others closely associated with the abortion of the unborn child. In such cases, supporting abortion could, indirectly, result in the creation of a society inundated and pregnant with a number of mentally unstable individuals. Such conditions not only can affect individual family structures but the general society as a whole. There could also be the argument that to allow young people to resort to abortion could lead to a lowering of personal values, morals, and standards on their part as well as giving the perception that life is not important. This could in turn lead them to be promiscuous and even become young murderers in other ways, such as killing with a gun or other weapons. Finally, there is also the belief by some doctors that abortion may lead to some kinds of cancers such as cervical cancer in women who have done abortion, especially more than once.

I have attempted to shed some light on the question of abortion. Several possible reasons that may be proffered for its support and application have been discussed. In the same light, several points that may be brought to the fore against abortion have also been presented. It is now up to you, my readers, to take your position. I would humbly advise that you continue to read as much as you possibly can on the subject. Further, I implore you to continue the discussions, with friends, read science journals and attend fora (forums) organized to this end, with a view to have a broader and clearer comprehension on the issue of abortion. It is important that you keep informed, participate in discussions and debate, and let your opinions be heard. It is in reading materials like this one, that your knowledge and confidence will be improved, making you more prepared for our gradually, ever increasing world of complexities, particularly in the twenty-first (21st) century!

TEST YOUR LEARNING

---

(1) Using information from this essay, additional resources such as your local and/or school library, (a) explain whether the author's presentation included any bias or biases? (b) Would you agree that the author's approach to the question of abortion is straightforward, concise, and easily digested? (c) Were you impressed, or not, of his seemingly novel approach to the question of abortion? Explain in each case.

(2) Would you agree to, or support those arguments furnished by the author with respect

to abortion as reasonable, and ones that might have been offered by groups speaking for, or against, abortion? Explain.

(3) Using information from this essay, additional resources such as your local and/or school library or by interviewing an expert, (a) did the author adequately and convincingly present his arguments in support of his views? (b) Do you think his points were mostly original? (c) Could the premise of his arguments stand up to critical scrutiny? Explain in each case.

(4) On a scale of 1 to 10, 10 being the best, 1 the worst, (a) how would you rate this essay? (b How would you rate his introduction to the reading using the same scale? Explain. (c Are there any supporting details that the author provides, that are not relevant to the discussion or in making his point? (d Could he have provided more supporting details to strengthen his arguments on both sides? If yes, what arguments or points could he have included? (e Could he have supplied more opposing points of view?

(5 Did the essay change your opinion on the abortion issue? If so, in what way did it change your view on the subject? If it did not change your opinion on the matter, explain why not.

ESSAY SEVEN

WHAT IS THE SIGNIFICANCE OF ETHNIC POLITICS IN AFRICA?

Ethnicity in Africa has been called the "resilient paradigm." What accounts for the ongoing salience of ethnic politics in Africa? Can it be accounted for using received explanations of its significance? How can ethnicity be linked to discussions of other social movements in civil society?

It is highly inconceivable that any learned individual, group or institution would deny the fact that ethnicity and tribal attachments or ethnic proclivities bear tremendous salience to every form of human existence throughout the length and breadth of this planet, including Africa. The central discussions of this paper will aim to make explicit the fundamental, underlying reasons that ethnicity offers such a significant impact and role in Africa's political systems. It will, in general, argue against the traditional rhetoric, "being verbiaged" by the West for its crucial and inevitable role and significance in Africa's politics. Finally, the interconnectedness with ethnicity and social movement in civil society will be illustrated in this essay.

It is probably prudent from the outset of this discussion to establish some general statements as to the meanings of certain key terminologies that are employed in this paper. The term or phrase, "resilient paradigm" connotes or signifies some kind of return to, or moving back to one's original form. It implies, as used in the question to this paper, some measure of continuity with ethnicity playing some intertwining role in the anatomy and physiology of Africa's political systems and structures. Secondly, the terms ethnicity and tribe will be used somewhat interchangeably though, in practice, the latter connotes some deeper, more primitive relations with groups of peoples, usually as perceived by the West. The terms, as will be used in this essay, signify a social group having a common cultural tradition and whose members have a common origin by birth or descent.

The permanent and eternal significance of ethnic politics in Africa has several explanations. Firstly, the complexities of relations and moral dimensions in Africa are rooted in, and emanated from, culture and ethnicity. The reality and history of material deprivation have forced peoples who share much in common to isolate themselves in strata or groups with a view to consolidate the not-too-accessible skills and resources they possess. This in effect leads to conflict with outgroups and to the creation and facilitation of tribal factions. Secondly, and closely related to the first explanation, ethnicity and ethnocultural diversity are utilized as a springboard for the ruling class to attain their political ambitions. The oligarchy exploits the notion of ethnicity to retain political power and the status quo. This is effected by leaders of dominant ethnic groups exploiting their members' lineage of their kinds, for their own selfish interests. In this stead, ethnicity is essential for political advancement of the few, to the domination and deprivation of the many. Put another way, ethnicity is significant in the political arena as it serves as an accessible vehicle for the rich and the powerful to cling to the elm of political manipulation, overpowering the masses of depressed African peoples!

There are other ways in which the ongoing salience of ethnic politics in Africa can be accounted. Firstly, a possible explanation is the fact that even though, in principle, the African peoples have been liberated from the shackles of colonialism and formal exploitation, there still obtain the informal and alien forces of imperialism. It is, therefore, not farfetched for one to imagine and for this writer to argue that to continue and proliferate the political status quo, it is vital that the African elite, from a particular ethnic, opportunistic background, be allowed to remain in force. What more effective way is there to establish this political control but through ethnic relations?! One's political future, in this respect, is closely related to how effective ethnic and tribal groups can be manipulated. Ethnicity's salience in Africa's politics is crucial on the basis described above.

Patricia Stamp in her article, *Burying Otieno*, also implies one of the reasons for the ongoing salience of ethnic politics in Africa. Patricia Stamp (p2) argues that there exists a great disequilibrium in gender relations in Africa and this is, in turn, closely linked to the structures and origin of Africa's political economy, among other African systems. She pointed out that Africa's political agenda is one which is based on, and intertwined between, male dominated kin groups and the state, and these serve to maintain the status quo of patrimonial domination and the enhancement and promotion of female subservience (paraphrased). Ethnic politics in Africa, in this respect, becomes crucially important as a significant number of ethnic and tribal groups view their women membership as less than their equals in ability, strength, status, and other forms. This ethnic relationship with Africa's politics, this union, serves and provides the political structure with an instrument by which it can subtly promote its repression and discrimination against women.

Again, Stamp can be quoted. According to her (p2), "The state is not a monolithic entity but a contradictory, disunified set of structures, processes and discourses."

To my mind, there is no doubt that ethnicity and its relation to politics are part of these structures, processes, and discourses of which Stamp, so eloquently speaks.

There still exist other and additional explanations for the ongoing salience of ethnic politics in Africa. Politics is concerned with the governance and the maintaining of control over a group of people in a particular geographic location. Ethnicity, therefore, becomes extremely essential in Africa's politics to the extent that Africa is founded and survived on cultural and ethnic diversity. It is through ethnic and cultural means, the argument is, that the African masses will gain some kind of understanding of its political processes but equally, it is with the guise that political benefits are best distributed through such existing structures, that ethnic politics engages this crucial and significant form. Ethnic politics is believed to offer and serve as the basis upon which the masses of the African peoples can be made to incorporate their way of life into productive participatory play. How practical and real this is in Africa, is a totally different matter. It is commonly believed that this state of affairs will bequeath upon Africans that of which they have been denied for ages. For instance, ethnic politics can serve both positive and negative ends.

Mlama points out in her book, *Culture, Politics and Development*, this very point. In chapters one and four of her book, she argues that on the one hand ethnicity is used as a vehicle for genuine education and the promotion of a deeper understanding of the African way of life as it relates to production, the economy, and politics. On the other hand, she discusses the way ethnic and cultural affiliations and affinity to tribal groups are used by the powers-that-be to carry out a political agenda in line with the elite classes and the colonial god-fathers and imperialists, to the detriment of the common African peoples. It becomes quite clear then why ethnic politics becomes so crucial in Africa. If politics is to be used as a vehicle of development and democracy in Africa, and ethnicity is so intimately linked with it, it is necessarily crucial for such a state of affairs and arguably, this is one of the bases of its significance!

Ethnicity and ethnic politics in Africa by necessity, are bound, if compelled, to afford and inculcate some of the values required in the local development of the marginalized poor. One such evidence is gleaned through the use of popular theatre in facilitating deeper understanding of its political, economic, and productive sectors. This is attained not merely through the sensitizing role it plays to the ordinary peoples but also in the enhancement and promotion of their development and potentialities.

Ethnicity is the "resilient paradigm" as mentioned above, as it continues to be an important link in the chain of Africa's political life. It is also important to point out that ethnic affinity or cultural identity and politics may form a symbiotic relationship. Ethnicity can be made to be a natural component of Africa's development by utilizing resources that can effect and foster development and economic growth. It is also imperative to indicate that in certain respects a parasitic relation may evolve as well. When political elite and the upper echelons infiltrate and exploit ethnic groups for their own selfish ends, this is exactly the kind of parasitic relation that obtains. The rich gets nastily richer whilst the marginalized, financially handicapped, and impoverished masses are further reduced to the seams of their rags!

Ethnic politics in Africa also becomes important for other reasons. Politicians may take advantage of an ethnic group's ignorance and shortcomings to advance their political agenda. It is therefore another significant reason that ethnic politics in Africa continues to be the "resilient paradigm."

Mhlaba states in *The Local Cultures and Development in Zimbabwe: The Case of Matabeleland* (p3) "Until you can rediscover your identity, until you can have confidence in your own capabilities, which have been proven for certain epoch, you will be dominated."

The ongoing salience of ethnic politics in Africa is imperative for a number of salient reasons other than those discussed in the foregoing pages. T. O. Beidelman in *The Kaguru: A Matrilineal People of East Africa* (p80), had this to say, "Where the law and political institutions are not closely enmeshed with other sectors of social life, we may expect a rise both in repressive actions by those in power and in conflict following attempts by those below to elude or manipulate unacceptable rules through illegal or extralegal means."

Beidelman is forthrightly and correctly implying that in the case of Kagaru, and I would argue by extension, Africa as a whole, there can be no stable political structure without considering ethnicity and culture. Having a political structure in place without consideration for ethnic and cultural background is analogous to an individual attempting to fix a broken computer with neither cognizance of its intricate components nor the identification of its problems! In this regard ethnicity becomes significant for the induction, promotion, and facilitation of any kind of genuine growth and development for a people. Another important point is that the political system of Africa needs to incorporate the rich experiences, knowledge, and way of life of Africans in enacting legislation. This is extremely crucial in that not conferring with the common, ordinary people could result in resistance to laws that are conflicting and contradictory to the lifestyles, values, and morals of these ethnic groups. Put another way, ethnic politics, in addition to being the most suitable route to Africa's development, is vital in the prevention of laws, inter alia, that are inimical, counteractive, and counterproductive to those essential elements of the lives of ethnic groups.

T. O. Beidelman (p81) made this statement, "When I lived in Kagaru, many government policies were determined not by the Kagaru people within their homeland but by the Europeans residing outside that area, some in the Colonial Administrative Centres and some back in Europe."

The foregoing quote speaks volumes about the state of affairs as they exist in Africa's politics in general and the role and participation of its peoples in particular. It serves as a wanton testimony to the scant regard and exclusions of the African peoples in determining their own destiny. It reveals the outright indifference and disregard shown to Africans as intelligent members of homo sapiens!

The importance of ethnic politics in Africa continues to wave its attractive head. Consider the situation in Kagaru just a few years ago. The British had to approve the election of their headman or chief and could remove him if they felt an urge to, whenever they were so pleased. Further to these catastrophic situations, was the fact that these chiefs were usually chosen from a certain ethnic group to implement and effect the whims and fancies of the alien nation to the exclusion and isolation of the dreams and goals of the masses! These Europeans, incognizant of the depth of local affairs, relied on the chief's assessments as a basis for the implementation of Kagaru's policies. Usually, the semblance of the official model could be preserved because of the chasm in communication between various strata of governments. The local leaders often misrepresented the truth in order "to look good" in the eyes of the European powers. At the same time, manipulations were being effected in various levels of European administration as well. This was with a view to impress their superiors. In Kagaru, for example, during colonialism, local government intensified tribalism to intimidate tribal minorities in their midst and at the same time reduced any significant chances to dissent and resistance. The Kagaru Native Authority, an organization created along ethnic lines, was used to foster special ethnic groups to the exclusions of others. This depicts the conservative outlook of the local government which was their way of guaranteeing the continued support of Britain. In other words, the local leaders of Kagaru ensured, by living up to the rules of the alien nation that they continued to receive a flowing stream of support to ensure that their political reservoir would never be void of rich fluid for their own selfish ends.

Ethnicity sometimes had effective impact in the economic and political direction of Africa. In kenya, during the early twentieth (20th) century, different peoples of various ethnic and cultural backgrounds, got together to resist exploitation from European powers. The Kikuyu tribes formed alliance with other ethnic groups to vent resentment in political expressions. For example, the Kikuyu Association Alliance was founded to this end. In Nairobi, in particular, there were different ethnic groups in alliance. Among those was one comprising Luo and Luyia and members of other ethnic backgrounds or kinship.

To quote Robert July from *A History of the African People* (p502), "In Senegal, there was great shortfall in the hopes for reform toward democratic self-determination and the crushing reactions of post war French policy."

That which Africans seemed to have been saying, in this respect, was that there was greater strength in numbers and the utilization of skills and varying experiences, was required to achieve a political governance in accord with the goals and aspirations of the peoples of Africa in general, different and apart from the selfish goals of the elite and the external forces of the West.

Finally, ethnicity in Africa's politics is crucial as it is a fundamental structure in the

African society. It is likened to the foundation of a high-rise building. Without a firm base the pull of gravity will soon have it in worthless shambles! So it is, with an Africa whose political directorates attempted to govern without taking into account the varying ethnic backgrounds and that which was uniquely associated with those experiences.

Colin M. Turnbull, in *Tradition and Change in African Tribal Life*, (p195) states: "Tribe and ethnic consciousness are very important . . . African society, by means of its closely knit family structure, fosters ethnic consciousness."

It is clear, that ethnicity and tribes serve to foster continuity and avoid breakages with the family system.

Ethnicity in Africa's politics as the resilient paradigm cannot be accounted for on the basis of the usual received explanations. Firstly, the West presents ethnic traditions in Africa as some uncivilized, backward phenomena which have no place in modern so-called civilized society. This, of course, is ridiculous nonsense. The cultural and ethnic traditions of Africa evade(d) the imperialists' comprehension, and consequently, is one explanation for this abhorrence on their part. Secondly, as cited earlier in this paper, alien forces have an objective and interests in Africa for their own selfish aggrandizement and accordingly, is set on painting a picture of doom and gloom for cultural and ethnic traditions. Thirdly, local elites, mobilized by the alien forces, at times pit ethnic groups against each other with a view to decry Africa's ethnicity as divisive and a liability to true growth and development for its nation states.

Ethnicity can be connected to discussions of other social movements in civil society in a number of ways. One way in which ethnicity may be discussed in civil society is in the situation which arises as a result of industrialization such as that of South Africa. As is discussed by Mamdani, *Citizen and Subject: Introduction* (p8), "As the economy of South Africa became industrialized, it gave rise to "colour problem", at the root of which were urbanized or detribalized natives."

Industrialization leads to the disintegration of many ethnic groups as individual members of families move away to various cities to "eek" out a living. This process was a sure sign of establishing the abandonment of the native tribes of any nation. Civil society with its implementation of a single legal order, lends itself to the repression, reduction, and eradication of ethnic and native existence. It dictates and nurtures the non-recognition of ethnic and cultural practices in exchange for conformity to European ways of life and legislation. Put another way, alien society has no room or patience for Africa's ethnicity as this is considered to be obsolete and primitive as far as Europeans are concerned. For ethnicity to survive in this segregation of social standards in civil society, it means forgetting one's roots and assuming a new, non-African outlook. In short, one is obliged to turn her or his back on all that one stands for including any recognition of lineage to those members of his or her tribe who have not yet been converted to the alien mode of life!

To quote Mamdani (p17), "Citizenship would be a privilege of the civilized; the uncivilized would be subject to an all-round tutelage. They may have a modicum of civil rights, but not political rights, for a propertied franchise separated the civilized from the uncivilized."

In other words, to be recognized, members of ethnic and cultural Africa must be inducted and converted to the norm and values of these aliens to be regarded as worthy and only then may they be likely to enjoy the benefits of so-called civilized peoples!

Ethnicity can also be linked with other social movements than those alluded to, in the aforementioned paragraphs. One such connection can be made with respect to the role ethnicity or tribalism plays in the colonial state.

As Mandani states (p24), "Everywhere, the local apparatus of the colonial state was organized either on an ethnic or a religious basis."

He continues to state that for most peasant uprisings in Africa, it was extremely difficult not to be able to find some kind of ethnic or religious inspiration in its occurrence. Ethnicity was, in most cases, recognized as problematic in every peasant's movement.

Finally, ethnicity is affected by new legislation which made a fundamental and revolutionary change to customary laws. In significant numbers of African states, customary rules which are particular to each tribe are replaced and discarded for a single customary law cutting across all ethnic boundaries. Its consequence is the development of a unitary and uniform customary law for the rural parts which is applicable to all peasants irrespective of ethnic backgrounds. These customary laws run simultaneously with modern law, applicable to urban ethnic groups. These changes, however, have negligible benefits, if any, to the masses. If nothing, it deepens the struggles of Africans. For one thing, there now obtains through these changes, a great chasm between peoples, the urbanites and the ruralites. This, in itself, deepens any existing conflicts between individuals of different ethnic backgrounds. In addition, the economic pressure of an already poverty-stricken masses was further intensified through these implementations.

This essay has shown that ethnicity in Africa is extremely significant for a number of important reasons. It has shown that ethnicity must be considered in the political, social, and economic contexts of Africa. It has also been revealed that the explanations usually given for its salience must be taken "with a grain of salt". Finally, it was shown that ethnicity can be related to a number of social movements in civil society.

TEST YOUR LEARNING

---

(1) Using information from this essay, additional resources such as your local and/or school library, (a) explain whether the author's presentation included any bias or biases? (b) Would you agree that the author's approach to the role of Africa's ethnicity in that continent is straightforward, concise, and easily digested? (c) Were you impressed, or not, of his seemingly intellectual approach to the question of the

significance of ethnicity in Africa's development? Explain in each case.

(2) Would you agree to, or support those arguments furnished by the author with respect to the role of Africa's ethnicity as being reasonable, and ones that might have been offered by expert historian groups speaking to the same topic? Explain.

(3) Using information from this essay, additional resources such as your local and/or school library or by interviewing an expert, (a) did the author adequately and convincingly present his arguments in support of his views? (b) Do you think his points were mostly original? (c) Could the premise of his arguments stand up to critical scrutiny? Explain in each case.

(4) On a scale of 1 to 10, 10 being the best, 1 the worst, (a) how would you rate this essay? (b) How would you rate his introduction to the reading using the same scale? Explain. (c) Are there any supporting details that the author provides, that are not relevant to the discussion or in making his point? (d) Could he have provided more supporting details to strengthen his arguments in the presentations of his points? If yes, what arguments or points could he have included? (e) Could he have supplied more possible contrary points of view to show balance in his presented arguments? Explain.

(5) Did the essay change your opinion on any prior perception(s) you have of Africa? If so, in what way did it change your view on the subject? If it did not change your opinion on the matter, explain why not.

ESSAY EIGHT

ARE LAWS THE SAME TO EVERYONE? WHO AND WHAT DETERMINE LAW?

Federal Government of Canada Enacted Statute: "Henceforth, all white males are excluded from employment by any branch of government or any organization, enterprise, business, or institution which receives government funding.

This paper seeks to establish whether the aforementioned statute would be law from the perspectives of three different legal theories. It will investigate and reference the theories of Aquinas, John Austin, and H. L. A. Hart, respectively throughout the presentation of his arguments. The writer will argue against the statute being law and in so doing, not only will he base his discussion on Aquinas's pronouncements, but he will discuss his own subjective philosophy on the issue.

There are several implicit conditions in the preceding statute which would deem it "nonlaw" from Aquinas' perspective. Firstly, the statute by its very nature is prejudicial, demoting and frustrating the wellbeing of one group of individuals to the apparent advancement of the other. It is therefore inhibiting to the common good which would not result in very compatible "bed fellows" with one of Aquinas' fundamental conditions for law. That being all law must be directed to the common good to be considered as such. Secondly, and closely related to the first argument, is the fact that the statute does not foster the common interests of the whole community. It, to the contrary, stifles the growth and self-actualization of a particular group which, by nature, deems it immoral. Consequently, from Aquinas' view point the statute is not and could not ever be law! As is common knowledge to all readers of Aquinas' Natural Law Theory, law and morality are permanent, inseparable, Siamese twins. Where one is, there exists the other. To elaborate, morality is the very being of law and vice versa. The existence and survival of law depend on this symbiotic relationship between the two. Thirdly, the statute fails one of Aquinas' most salient tests. It does not seem to emanate from reason. To the extent that it discriminates so blatantly against one group, means it could result in serious repercussions for the society. Such implications, inter alia, include starvation of this group as a result of pecuniary difficulties emanating from unemployment. The statute also lends itself to the development of grave factions within the society. This of course could lead to civil disturbances and instability among its peoples. All these and the other debilitating factors, too numerous for inclusion in a paper of this scope, would depict some kind of shortsightedness on the part of the legislators and in tandem some lack of reason on its part. Any statute which has the potential for the kind of instability and possibly, anarchy as described above, appears to be shortsighted and deficient in good reasoning. Clearly, for any pronouncement to be law, according to Aquinas, it must be directed by reason.

Several other factors implicit within the alluded to statute make it "nonlaw" according to Aquinas. The statute though being promulgated by the government, contradicts one of the most vital principles of Aquinas. It is not in line with the doctrine of God. For any statute to be law, he argued, it must be guided and instructed by the principles of God. This "God principle doctrine" is, to a significant extent, Christian based. Those of us who have been exposed to Christian teachings (whether we are seriously engaged in such matters or not) will, undoubtedly, have no qualms that the statute alluded to is not harmonious and in accord with Christian ideology. As the statute is not "God influenced or motivated", it stands to reason that it does not emanate from Eternal Law which for him can be the only origin of law. The statute also promotes inequality and injustices on the part of the victims and necessarily breeds immorality. In light of the foregoing, the statute is clearly not law!

Notwithstanding the aforementioned conclusion that the alluded to statute is not law, other legal theorists would argue to the contrary and would no doubt speak voraciously in favour of the statute being law. One such theorist who would argue this is Austin in his Positivist Law Theory. Austin, unquestionably, would agree that the statute is law for a number of reasons. The first argument to support the statute as law is maintained by the Positivists' philosophy. That is, law is law and has nothing to do with morality. For Austin and some other Positivists, morality and God's teachings are separate and apart from law and consequently serve as no "yardstick" in determining what law is. Secondly, Austin would argue that the statute is a clear command which has implicitly some kind of sanction or penalty to be incurred upon the party to whom it is directed in case it is disobeyed. As a consequence of the inherent content and nature of the statute, there exists a duty or obligation to obey it which qualifies it as law. The fact that this statute was enacted by the Federal Government of Canada makes even more forthright the point that this statute is in fact law. His discussion would be based upon the notion that the enactment emanates from the nation's Sovereign, the legal instrument for the enactment of laws. The Sovereign, in Austin's view, has this unique power endowed upon it by reason that the majority of the given society is in the habit of obedience or submission to it while the Federal Government, Austin would argue, is not in the habit of subservience to the group to which the law is directed.

There still exist other viewpoints that would support the statute as law from Austin's Law Theory perspective. The fact that its enactment is through the Federal Government of Canada, to a great extent should qualify it as law having been produced by a rational being or a group of rational beings. It will be assumed that the term rational connotes some kind of reasonable, intelligent homo sapien or groups thereof, as discussed by Austin. The mere fact that Canada is recognized internationally as a civilized and democratic society, speaks volumes to the fact that its sovereign, in this case, the Federal Government, is rational and accordingly, this statute must be law. The second point is that the statute is not occasional or particular in nature but is a command that obliges one generally to an act. Put another way, the statute commands that effective immediately, a certain legal requirement in the employment process be established. This qualifies as law as the required behaviour is not one limited to a particular period or temporary time frame, even though it obliges exclusively persons individually determined, the male Caucasians. Finally, in support of the statute as law, it would be argued that in light of the fact that the Federal Government satisfies the conditions required, as presented above, and inasmuch as it has political power as sovereignty to offer some kind of sanction, evil, or punishment, the statute thus enacted can be nothing less than law!

Inasmuch as from the perspectives of the previously discussed theories, a declarative, straightforward, pronouncement was distinctly made as to whether the statute was law or not, in Hart's case, this relationship to law is contingent on certain factors. In tandem with Hart's Positivist Theory, the statute alluded to in this paper, cannot be automatically considered law merely because it is enacted by the sovereign's coercive orders. Such a statute, for Hart, could only be legally binding if the persons to whom it is directed perceive themselves as having an obligation to abide by this statute. For Hart, having an obligation or duty is defined differently from Austin. The former, unlike the latter, conceptualizes "having an obligation" as subjective in nature and is not so much related to whether there exist or not some kind of sanctions or penalties for breaking an order. Therefore, to declare whether Hart's Theory would make this statute law, a closer analysis of it must be made. The context of this scenario needs to be examined. The question that needs to be asked is, does the aforementioned statute lend itself to the majority concerned, obeying or rejecting it? Put another way, is it likely to be agreeable to the majority to whom it is concerned? It is in correctly answering this important question that one can positively state whether the statute is law or some tyranic, onesided verbiage. One could reasonably assume that the discriminatory nature of the statute would elicit universal condemnation among those to whom it is directed. In this regard, the statute would not be law. This is to say, that those who the statute affect, do not feel any obligation to obey it. In other words, the nature of the statute does not breed or if one prefers, elicit some kind of social pressure or social threat such as guilt, shame or remorse. This internal point of view is relevant as to whether the statute is law or not. On the other hand, if the majority concerned found it agreeable and the individuals violating the statute feel pressured to conform, not conforming would involve violating one's obligation. From this standpoint, the statute is legally binding and is law. There is absolutely no doubt in this writer's mind that whether the alluded to statute would be law or not, from Hart's perspective, would be contingent on the foregoing discourse presented. The statute would therefore be law only on the basis that it brings pressure to bear on those who deviate and act contrary to its commands. If it so engages in this regard, the statute would be law.

It is this writer's view that the statute of reference if anything, is divisive, wantonly biased but a far cry from being law. The very nature of law, as Aquinas would agree, makes it practically impossible to be severed from morality. To the extent that a cabinet minister brings before the House a bill that requires doctors to shoot on spot all her HIV positive patients on discovery of the condition, would elicit shame and condemnation from all but psychotics, equally, a statute of the dimensions discussed here, must be rejected on its immoral nature! Discrimination is discrimination irrespective of the form, shape, or colour. The mere fact that even just reading this analogy evokes such absurdity and seems to be farfetched, speaks even more powerfully to the disdain of this statute. One's perception as to the purpose of law will undoubtedly vary, but the notion that among the objectives of law, as seems to be implied by this statute, is to create physical and psychological pain to ordinary, law-abiding citizens cannot, by any form, be accepted by sane, civilized individuals!

It is my view that Aquinas' work best addresses the argument as to the legality to such a statute although some possible reasons that he might have based his arguments on would not be considered by me. As can be gleaned from my earlier arguments, it is a fact that laws in most cases must be based on moral principles. This is to say, it must be founded upon that which is considered right or wrong for a given society. On the contrary, it is not this writer's view that laws have to be guided by God's principle. To suggest, for example, that the statute of reference is not law solely because it is not of God is to embark on dangerous grounds. It would have to mean that God and God's teachings are one and the same to all peoples. This in reality is not the case and there are, in fact, religions whose God would probably make the statute being discussed law! As expounded in Aquinas' Theory, it is imperative that laws evolve from reason. This statute was deficient in this respect as explained before. It is also obvious that the statute's end was not for the common good. Another factor that makes this statute fall short of true law is that the statute did not have an equal impact on the community, as a whole, but was directed to a special group of people. This, of course, makes it morally chaste and in effect "non law".

In conclusion, it must be noted that it is practically impossible to separate law and morality. This fact is even more in your face when one analyzes laws that have evolved from countries that are so-called civilized, modern, and democratic societies. And even in situations where law portrays some semblance of semi-immorality, the law usually is not enforced by the law enforcers or applied in the case of judges. For instance, studies have shown time and time again that the harsher the sanctions (and one could argue the more immoral and unjust it is), for violating a law, the less likely it is to be enforced. For example, in Jamaica where the penalty for committing murder is the gallows, fewer people accused of murder have been actually found guilty of the crime compared to countries where capital punishment is not enforced. This depicts how closely law and morality are enmeshed in practical terms even if denied on a theoretical basis. It is not very difficult for you to imagine the following situation: The Canadian legislature makes it punishable by death all drivers caught speeding. It is not farfetched to conclude that the law enforcers would be very reluctant to make an arrest in this regard. This kind of legislation would be very rarely enforced if any at all. Why should this be?! It would be so because most law enforcers would perceive the law as unjust or immoral. It is for these, and the abovementioned reasons, that this writer's contention continues to be that the aforementioned statute would not and could not operate as law.

TEST YOUR LEARNING

(1) Using information from the essay, additional resources such as your local and/or school library, (a) explain whether the author's presentation adequately and correctly reflects the views of Hart, Austin and Aquinas? (b) Were you impressed, or not, in the way he attends to the theories to make his points? Explain in each case.

(2) Would you agree to, or support those arguments furnished by the author with respect to the circumstances under which the given statute would be considered law or not? Explain.

(3) Using information from this essay, additional resources such as your local and/or school library, or by interviewing an expert, (a) did the author adequately and convincingly present his arguments in support of his views? (b) Do you think his points were mostly original? (c) Could the premise of his arguments stand up to critical scrutiny? Explain in each case.

(4) On a scale of 1 to 10, 10 being the best, 1 the worst, (a) how would you rate this essay? (b) How would you rate his introduction to the reading using the same scale? Explain. (c) Are

there any supporting details that the author provides, that were not relevant to the discussion or in making his point? (d) Could he have provided more supporting details to strengthen his arguments in the presentations of his points? If yes, what arguments or points could he have included? (e) Could he have supplied more possible contrary points of view to show balance in his presented arguments?

(5) Did the essay change your opinion or any perception(s) you have about law? If so, in what way did it change your view on the subject? If it did not change your opinion on the matter, explain why not.

ESSAY NINE

IS IT UNCONSTITUTIONAL TO IMPOSE LAWS AGAINST THE ACT OF SODOMY IN CANADA?

Would the following law be justified in Canada today?

"Any person who engages in sodomy is guilty of an offence. Sodomy for the purpose of this act shall be defined as any sexual intercourse between members of the same sex, or any acts of oral or anal sex, whether between persons of the same sex or opposite sex. The penalty for anyone found guilty of the offence under this law, shall be imprisonment for not more than five years and not less than one year."

This paper seeks to establish whether the above law would be justified in Canada today from the perspectives of different theories and philosophical writings. To effect such an analysis, I will engage the theories and writings of John Stuart Mill, on Liberty, Dworkin's Paternalism Theory, facts from the *Bowers v Hardwick* case of the United States and excerpts from the *Canadian Charter of Rights and Freedom*. I will argue against the justifiability of such law in Canada today.

There are several factors inherent in the above law which would deem it unjustifiable from Mill's perspective. Firstly, and very significantly, for him, Mill's outrage would be expressed and vented on the basis that the law sets out to admonish and punish an act that is self-regarding. Put another way, one's decision to engage in a sexual act, as is described in this law, though not encroaching on the interests, prejudicially or otherwise, of others, is made to be a crime. According to Mill, the only correct basis on which one's liberty can be curtailed in any way is with respect to the "Harm Principle". The "Harm Principle" stipulates that the liberty of others may only be restricted on the basis that such liberty will cause harm to others.

"What harm can the acts of homosexuals or sodomites cause to others?" might be Mill's query.

Secondly, Mill would argue that even if the act of sodomy in one's private domain could be proven somewhat inimical and offensive to others, there exist various nonpunitive ways to address the matter. If one finds the act so terribly obnoxious, Mill would argue that instead of declaring it criminal and justified to be punished, alternate steps could be made to elicit a desired outcome. Society could employ positive measures such as education and persuasion to attain that end instead of interfering in one's liberty to do as she or he pleases with respect to one's private life, by punitive means.

The third factor which makes the Sodomy Law unjustifiable from Mill's point of view is that it denies and stifles one's independence.

Mill states, "The only part of the conduct of anyone for which he is amenable to society, is that which concerns others. In the part which merely concerns himself, his independence is of right, absolute."

There is no doubt that the aforementioned law infringes on one's independence and autonomy in deciding one's sexual preferences. It is clear that this infringement has nothing whatsoever to do as an intervention to prevent harm to others. This, of course, would be frowned upon by Mill and the law would be declared unjustifiable!

The law becomes unjustifiable on other grounds as dictated in Mill's Liberty Theory. The fact that Canada is a developed, civilized society makes such a law even more wanton and unacceptable and cannot be recognized on any moral basis. The Theory postulates that all mature persons of civilized societies must be allowed to choose their own engagements as they see fit provided, they do not interfere with those of others. The Sodomy Law, here described, cannot be therefore justified on the grounds which are considered as the best action for those to whom it is directed. It denies and deprives one's right to choose. Even if this law is a reflection of society's majority and bearing in mind the curtailment of one's liberty as discussed above, the law still would be in conflict with the Liberty Theory of Mill and on no grounds, morally or otherwise, could it be law in Canada today, according to Mill's Liberty Theory.

Notwithstanding the above argument that the Sodomy Law is unjustified in Canada today, other moral theories may argue to the contrary. One such theorist who may argue that the law alluded to is justifiable is Gerald Dworkin in his Paternalistic Theory. There may be considerable reasons that may be posited in this regard. Firstly, it may be argued that in light of the severe discrimination meted out to homosexuals and similar kinds, the law is necessary to protect individuals from engagements in such activities as a means of promoting their interests and welfare. It may declare that one's economic wellbeing may be significantly retarded through discrimination in employment practices, leaving the individuals with very little to enjoy a satisfactory and productive life. In this regard, the argument may be, that the interference of one's liberty, as is the case with the alluded to law, is justified by reasons that it promotes, though indirectly, the welfare, good and interests of persons whose liberties are tempered. Secondly, the point may also be made that individuals who are engaged in sodomy may be ignorant of certain adverse consequences possibly associated with the sexual expression and consequently, need to be curbed by the hands of the law. The so-called adverse conditions may be listed as health risks, psychological trauma, physical discomfort, family conflicts, inter alia.

The potential adverse factors of health, psychological trauma, physical discomfort, and family conflicts usually advanced by non-sympathizers, may be presented as reasons to justify the Sodomy Law. The law may be considered as "morally good" and justified to the extent that it is considered to reduce the risk of certain transmittable sexual diseases, for example HIV/AIDS. It is commonly the belief that sexual activities which involve anal penetration are far more likely to transmit the HIV virus where at least one partner is HIV positive. On this basis, Paternalism will argue that law against sodomy is justified. It is justified as it is advancing the good health of the individual in limiting his liberty in this regard. It may be established that some cases of sodomy may be engaged in because of coercion, rather than one's will. Such coercions could take various forms. The need to gain acceptance in a particular group or it could be for pecuniary advancements. These cases could lead to psychological discomfort and on this basis, Paternalism would make the Sodomy Law justified. It does so, on the basis that such individuals are "saved" from psychological pain through this enacted statute against sodomy. Severe types of physical pain may result from the act, in particular to those individuals who are new recruits. For the same reason as stated earlier, Paternalism would again justify this Sodomy Law. Parents and partners who disapprove may also evoke family conflicts which in turn could trigger psychological difficulties in the persons engaged in the sexual acts. The Law of Sodomy in such cases would therefore be justified under the Theory of Paternalism.

There may yet be other arguments posited by Paternalism to justify the law. The likelihood that any psychological instability brought about by the act in some cases, may be irreversible or at best, difficult to eradicate could advance support for the law on sodomy. The speculation that young individuals may choose this form of sexual expression, without being fully cognizant of any stereotype and emotional content that may be associated with the so-called deviant, sexual practice, would support the Paternalistic trend. The Paternalistic Theory may raise support for the law in light of the foregoing and may lend significant justification to it.

Irrespective of the Paternalistic viewpoints in support of the statutes as being justified in Canada today, I tend to differ in that regard. I will present several arguments to establish the unjustifiability of such a law. My arguments and thoughts will substantially be original, but I will employ as well, arguments gleaned from the *Bowes v Hardwick* case in the United States. In this case, one Michael Hardwick, a homosexual, went to the Court of Appeals in order to be decriminalized by virtue of his sexual orientation running afoul of the Georgia Statute. The Georgia Statute was similarly worded to the law being discussed in this paper. The High Court reversed the decision of The Appeals Court that Hardwick should not be considered a criminal because of his sexual orientation. Those justices whose judgments in this case were contrary to the majority of the High Court, will also be used to argue against this law being justified. Arguments made by the Court in their decisions to reverse the Court of Appeals, will also be used to expose the flaws in them. The *Canadian Charter of Rights and Freedom* will also be used as a basis for my unrelenting view that the law discussed, herein, cannot be justified.

There are several strong arguments to prove and reveal the unjustifiability of such a law today. The first and most obvious, in my view, is the blatant violation of freedom of expression inherent in this statute. The fact that most "decent" law-abiding citizens engage in their sexual activities behind closed doors, out of the reach of public, speaks volumes to the unjustifiability of such a law in Canada today. Being in one's private domain, sodomy does not therefore affect adversely, or any at all, the life of the public. On this basis, there cannot be any moral basis to prohibit the act. *The Canadian Charter of Rights and Freedom* under Legal Rights, section seven (7) states:

"Everyone has the right to life, liberty and security of person and the right not to be deprived thereof, except in the principles of fundamental justice."

Obviously, one's sexual practices which are generally done in private, qualify for the freedom stipulated in this subsection of the *Canadian Charter of Rights and Freedom* and such a sexual practice, naturally does not frustrate any of the principles of fundamental justice. Secondly, and closely related to the first, is the question of privacy. The Canadian Constitution protects one's right to privacy. As discussed above, most individuals of sound minds are usually involved in the sexual act in some kind of private settings. To enforce such a Law of Sodomy would necessarily involve violating not merely one's right to privacy but also the rights to freedom of association and peaceful assembly as dictated under the *Fundamental Freedoms Principles* of the

Constitution. This law even delves further in the erosion and freedom of sexual practices of private individuals than might be recognizable at first glance. It seeks to restrain and make criminal the very methodology of sexual fulfillment between mature, consenting individuals. Put another way, its goal is to punish a harmless behaviour between private individuals because of some trivial, one-sided notion of society or group of influence. This, in itself, is very dangerous and cannot be accepted on moral grounds. To the extent that making it criminal to deviate from the norm of shaking hands in Western Societies to the kissing on the cheeks, and on the lips by others, from different cultures when greeting each other, would be scoffed at and viewed as insane, equally we cannot accept this law as morally justified. There is no constitutional basis on which the criminalization of either of these activities could be justly and legally effected!

In addition to the above arguments, other arguments can be stated against the justification of such a law. The law, as described above, deprives mature adults of the fundamental right to be left alone. Their rights to engage in desirable, private and consensual activities are frustrated by the law in question. This, of course, is unconstitutional. Secondly, as pointed out by those judges of the Appeals Court who differed with the decisions of the High

Court in the *Bowes v Hardwick* case, "Sexual intimacy is a sensitive key relationship of human existence, central to family life, community welfare, and the development of human personality."

There cannot rightfully be any moral basis to support the wanton, psychological traumas that such a law could bestow upon individuals for mere engagements in an act fully supported by the constitution by virtue of its privacy rights enactment. The statute also fails to recognize the significance of having the freedom to choose the content and style of the sexual act. The fact that individuals usually categorize themselves on the basis of their sexual relations, makes it even more imperative that this right, protected by the constitution by reason of its private nature should not be violated. The law also cannot be justified in light of the fact that the act of sodomy generally involves no intrusion or interference with others. Neither can it be rightly argued that the act of sodomy enforces its value system on others, as sexual encounters are usually done in the privacy of one's dwellings! And even if there are a few individuals who may not be as discreet in their acts of sodomy as the vast majority, this would also be true for sexual activities of non-sodomic nature. There still cannot be any fair, moral basis to deny the vast majority of sodomites of their rights, as all sexual acts would have to be condemned in this regard, to eliminate discrimination. It is common knowledge that some individuals occasionally partake in sexual activities in not very private settings. Do the legislators make the other "normal" sexual activities a crime on this basis? The answer is the negative. There can be no legal basis on which sodomites' rights can be denied them in this regard.

In conclusion, it is my view that the law alluded to in this paper, cannot be justified in Canada today. It cannot, in that, it violates the right to privacy, freedom of expression, and freedom of association enacted in the *Canadian Charter of Rights and Freedom*. Further, it fails to consider the very important role that sexual relations play in identifying and determining who the individual is as well as her or his personality. In effect, it sets out to contain and restrain the private lives of all sexually active individuals and accordingly, denies great numbers of individuals their rights to freedom of expression, right to privacy and association provided them by the Canadian Constitution. The Sodomy Law, it seems, sets out to violate one's constitutional rights, not because sodomy infringes on others' rights but apparently, merely because it is assumed that a majority would be in concurrence with such blatant and wanton piece of legislation! Would a statute, so enforced, as seems to be the law in question be justified? I think not! My opinion continues to be that the law alluded to in this essay, is a far cry from being justified and if anything, is wantonly absurd!

TEST YOUR LEARNING

---

(1) Using information from this essay, additional resources such as your local and/or school library, (a) explain whether the author's presentation adequately and correctly reflects the views of John Stuart Mill's Liberty Theory, Dworkin's Paternalism and the Bowes v Hardwick case in the USA? (b) Were you impressed, or not, in the way he attends to the theories to make his points? Explain in each case.

(2) Would you agree to, or support those arguments furnished by the author with respect to the circumstances under which the given statute would be considered law or not? Explain.

(3) Using information from the essay, additional resources such as your local and/or school library or by interviewing an expert, (a) did the author adequately and convincingly present his arguments in support of his views? (b) Do you think his points were mostly original? (c) Could

the premise of his arguments stand up to critical scrutiny? Explain in each case.

(4) On a scale of 1 to 10, 10 being the best, 1 the worst, (a) how would you rate this essay? (b) How would you rate his introduction to the reading, using the same scale? Explain. (c) Are there any supporting details that the author provides that were not relevant to the discussion or in making his point? (d) Could he have provided more supporting details to strengthen his arguments in the presentations of his points? If yes, what arguments or points could he have included? (e)Could he have supplied more possible contrary points of view to show balance in his presented arguments?

(5) Did the essay change your opinion or any perception(s) you have about sodomy? If so, in what way did it change your view on the subject? If it did not change your opinion on the matter, explain why not.

ESSAY TEN

SHOULD CULTURE BE CONSIDERED A REQUIRED COMPONENT FOR DEVELOPMENT IN AFRICA?

This paper seeks to describe and analyze the ways in which culture can be utilized as a component of development. To achieve this end, it will employ several arguments with a view to "hit the point home", so to speak. The writer will also demonstrate possible contradictions between development and culture as they pertain to Africa. It will also be established that culture has, in the past and even in the present, been utilized in ways beneficent to the ruling class and in the proliferation and promotion of the status quo.

There are fundamental and salient ways in which culture can serve as a component of development in Africa. Firstly, and very importantly, culture in the form of popular theatre, bequeaths upon the masses of the African peoples that of which they have been deprived for ages, the opportunity to incorporate their way of life into productive participatory play. This kind of indigenous and creative, playful engagement not only serves as a means of entertainment for local grassroot peoples but more significantly, as a vehicle of education. It facilitates insights into the masses and accordingly, initiates some semblance of true development. It, in so doing, fosters development of the local proletariat different from the usual pecuniary advancement of the elite and their imperialist powers! Secondly, and closely related to the first reason, culture, through popular theatre in Africa, promotes instead of disregards the local communication processes in Africa and accordingly, affords and inculcates the values necessary for local development of the marginalized masses. As these practices usually engage the wide masses of their society in researching their communities' needs, culture in this way serves not merely to sensitize the ordinary, common poor of their shortcomings but also the enhancement and promotion of their developmental growth. It does this by displacing the usual top-down processes designed and effected by the ruling class whose object was the mere satisfaction of their own interests!

There are other ways, apart from those discussed above, in which culture can be utilized as a component of development. An emphasis on the local cultures' needs equally and radically, signifies a revolution in the strategies for local development. It means a shift from the repetitive and constant considerations of the so-called economic and technical strategies of the powers-that-be to the expense of the majority's improvement. It, accordingly, obliterates the marginalization of the masses in this respect while encouraging real growth of the African peoples.

As is quoted in Mlama's book, *Culture and Development: The Popular Theatre Approach in Africa*, chapter one, Swantz states, "Not only a mentality of trust in one's own cultural heritage, but also a deep understanding of different cultural patterns and ways of perceiving and conceptualizing practical life situations is crucial for development in general."

It gives greater comprehension and insight into the elements necessary for growth among Africans. Culture provides the most effective vehicle in the arrival of Africans' developmental destiny as Swantz seems to agree. Culture can be made to be a natural component of Africa's development by employing anthropologists in effecting and evaluating potential projects for development. It can also be made to thrive through the establishment of financial support network for cultural identities and cultural projects. Culture, in order to be a component of development in Africa and other underdeveloped parts of the world, should desist from responding to the whims and fancies of the West and engage in the reflections and growth of endemic heritages of the local peoples. Culture serves as a component of development in that it leads to the awareness of the shackles which entrap them and provides the will and motivation to escape from subservience caused through the legacy of colonialism and imperialism. Thirdly, cultural tools such as education, religion, and language can be utilized to advance the understanding and rejection of alien economically promoted and motivated activities inimical to the real significant growth of the masses. Fourthly, because Africans have grown to believe that the capitalist way of life is the only route to civilization and development, culture in theatrical expressions can be used to uproot this false notion and can serve as a basis to liberation and development.

The ways in which culture can be employed as an integral part of development continues to be enormous. Through theatrical performances the correct and relevant messages for development can be conveyed without boredom and loss of interests. It does this as people are totally engulfed and engaged and consequently become self-motivating and motivated.

Luke Mhlaba states in his article, *Local Cultures and Development in Zimbabwe: The Case of Matabeleland,* "Cultural alienation renders difficult, even impossible, the effective mobilization of the masses for national development."

Culture, which expresses the life style, mode of production, strengths and weaknesses and the needs of Africans, serves as an extraordinarily important yardstick in evaluating and consequently, as a salient catalyst in the mobilization and inducement of development. It does this naturally through the opportunity that it offers for introspection of the masses by the masses themselves and not by some mere alien values which may be prone to great resistance. Resistance, in this fashion, leads to, and breeds, further erosion and degradation of any significant development of the African peoples. Cultural inclusions in Africa also promote development, in that it avoids apathy and lack of initiative. It is almost crystal clear that if a nation lacks initiative and is tremendously apathetic to its society's norms, development must be retarded in some significant way. In this way, utilizing the local cultural habits and forms can be employed to reverse and mitigate these psychological factors that stand in the way of significant growth and development of its peoples. Finally, culture leads to the understanding of who he or she is. To put it another way, emphasizing and working through one's culture lends itself to self-actualization and the discovery of one's full potential. What other way could be more relevant and effective in the real development of a nation?!

To quote, once again, Mhlaba, in the aforementioned article alluded to in this essay, "Until you can discover your cultural identity, until you can have confidence in your own capabilities, which have been proven for certain historical epoch, you will always be dominated."

It is clearly said and needs no elaboration.

There obtain possible contradictions between development and culture as they pertain to Africa. The ruling classes and the imperialist forces at work, in Africa firstly, have a notion of development differently conceptualized than the masses. Secondly, and in accordance with this concept of development, culture is manipulated to bring about an end which is in contradiction and conflict to the needs and aspirations of the ordinary peoples. So-called cultural engagements of the marginalized masses are sometimes used to advance the concept of development inimical to the realization of real and true growth. Culture, in the forms of theatrical engagements, is sometimes used by the ruling class to cajole, persuade, but more importantly, brainwash the masses in accepting the status quo. The nature of this tendency is so frequent and pervasive that it sometimes leads to organized resistance further leading to a retrograde step in Africa's development and the total frustration of the African people.

Cultural expressions and engagements are sometimes utilized in Africa in ways that are inimical to the masses' wellbeing but in proliferation and promotion of the status quo. It has been employed to effect development and democracy as conceptualized by the West. To some extent, the culture and "development" in Africa have been an induced and cajoled response to foreign interests and domination. The exclusion of the local culture is a deliberate means to create an economy dependent on, and subservient to, the minority ruling class and the imperialist forces of the West. In this social and economic design, it is ensured that the masses are kept in docile servitude while the minority continues to enjoy a significant amount of the means of production of the local peoples. It is a strategy designed to subjugate the masses to a feeling of inferiority to a superior minority who should be adored. It is a system which aims to acculturate a people to feel and exhibit gratefulness to a minority for even communicating and accepting favours from an undeserving majority.

Culture has also been in the advancement of foreign influences and the ruling class. African rulers have sometimes used culture, through theatre, to teach Africans to be grateful for what has been provided them by modernization. They have sought to instill in the masses that growth, development, and democracy must be attained by and through their prescribed strategies and that African cultural practices are, at best, backward and obsolete. It is forcefully communicated, through some of these theatrical practices and in other ways, that Africans must forget their roots in the name of development. This "development" and democracy are significantly directed at brainwashing and stealing the birthrights of the African peoples! This is illustrated in the raw poverty that exists in many African communities and the "voicelessness" of the majority.

As Mlama puts it, "African Governments censor the arts to allow only the art for arts sake and remove the others which may sensitize the masses to their exploitation."

Culture has come to signify and promote the ideology and philosophy of the legislators and the policies of political parties whose interests are incompatible with genuine growth of the African masses. This, in some cases, has led to significant numbers of Africans accepting and confusing the capitalist lifestyle with great growth and development and of exploitation, as success. Finally, several cultural tools (some of which have been already described in this paper) are used and operational in the fostering of a capitalist culture. The latter is designed to the support of the system in tandem with the ideology of capitalism and the capitalist structure.

In conclusion, it is this writer's view that culture, and in particular, African culture, forms an inseparable component of development. Culture and development are likened to Siamese twins. In such a condition exists a symbiotic relationship. Genuine development, for the African peoples, needs a thorough cultural foundation on which to build. Similarly, culture based on the roots and backbone of the common masses, is bound to be nourished by real localized development. In addition, it must be remembered that culture and cultural practices can be utilized to maintain the social structure of domination of the marginalized majority by the minority ruling class. It is with this awareness and knowledge that Africans must begin the embarking and revolution of real development, democracy, and growth. It is only with an understanding of the instruments at play and their role in the developmental game that one will be able to discern and identify the wolves in sheep's clothing. It is in understanding the forms, physiology, and anatomy of these wolves, however dubious, that culture will continue to form an integral part of development for Africans and not merely a brainwashing serpent!

TEST YOUR LEARNING

(1) Using information from this essay, additional resources such as your local and/or school library, (a) explain whether the author's presentation adequately and correctly reflects the views he proffered that culture is a necessary component for development in Africa. (b) Were you impressed, or not, in the way he lends support to his argument? Explain.

(2) Would you agree to, or support those arguments furnished by the author that the West and the local African minority ruling class have a deliberate and tailored agenda to maintain the status quo in Africa? Explain.

(3) Using information from the essay, additional resources such as your local and/or school library or by interviewing an expert historian, (a) did the author adequately and convincingly present his arguments in support of his views? (b) Do you think his points are mostly original? (c) Could the premise of his arguments stand up to critical scrutiny? Explain in each case.

(4) On a scale of 1 to 10, 10 being the best, 1 the worst, (a) how would you rate this essay? (b) How would you rate his introduction to the essay, using the same scale? Explain. (c) Are

there any supporting details that the author provides, that were not relevant to the discussion or in making his point? (d) Could he have provided more supporting details to strengthen his arguments in the presentations of his points? If yes, what arguments or points could he have included? (e)Could he have supplied more possible contrary points of view to show balance in his presented arguments?

(5) Did the essay change your opinion or any perception(s) you have with regards to the challenges Africa faces? If so, in what way did it change your view on the subject? If it did not change your opinion on the matter, explain why not.

ESSAY ELEVEN

HOW AND WHY WAS FORMAL EDUCATION INTRODUCED IN ONTARIO AND THE WEST IN GENERAL? WERE THE PUBLIC-SCHOOL PROMOTERS IN MID NINETEENTH CENTURY ONTARIO, MOTIVATED BY A DESIRE TO ADVANCE THE GENERAL GOOD OF SOCIETY, OR BY THEIR OWN SELFISH CLASS INTERESTS?

There were fundamental and radical changes with respect to educational activities and the educational system in mid nineteenth century Ontario and the West in general. In particular during the eighteen fifties, Ontario and Ontarians had witnessed a great inundation of the radical, if sweeping changes which had reached its highest heights before the turn of the new century. The question then is, what were the motives behind all the changes to education and the way in which they were engaged? Why is it that so many scholars and academics as well as politicians got all caught up in this mighty wind of change with regard to Ontario education in this era? Were these public-school lobbyists motivated to act for change as a means of self-aggrandizement and the proliferation of themselves and families, or were they unselfishly enhancing and augmenting the overall good of their society?

This paper will argue that for the most part, these education lobbyists were unselfish in the zeal to change education in the mid nineteenth century. It will do so by citing examples of the deeds and words of the more influential public-school promoters, among other methods. It will argue that the public-school promoters were inclined to their indulgence through a duality of sheer philosophical outlook, pervading the period in question, and the need to promote the common good of the society.

To give a fuller understanding to my thesis of unselfish need to promote the broader good of society and the philosophical view paraded on the school promoters' part, an insight will be given on the state of affairs with respect to educational activities prior to this period.

At this point, I will outline, briefly, the reason that my position in this essay is the correct one. That is, at this point, I will set out to elicit a deeper comprehension to my position and arguments in my readers' minds, and to make it more meaningful and convincing that it was significantly an urge for general societal advancement as well as a philosophical inclination on the part of the promoters, that this group of public schools lobbyist emanated. Firstly, the point must be made that prior to the mid nineteenth century, education, if it could be so described, was a loose, disorganized, insignificant factor in the lives of most individuals. There were not many individuals who regarded it as vitally important to survival until after the period earlier alluded to, and to the present day much interests, time and energy are employed in the acquisition of an education.

In the words of Alison Prentice (p15) in her text, *The School Promoters*, "Prior to the Ryerson era in Upper Canada and indeed, nearly everywhere in the Western World in the early nineteenth century, education was characteristically "voluntary" and informal. The usual and perhaps the fundamental educational institutions were the households, workshops, and fields since the vast majority of children learned most of what they needed to know from their parents, or from adults in other families to whom they were bound as servants or apprentices."

The Ryerson referred to in Prentice's quote speaks to one Reverend Egerton Ryerson who served as a Superintendent of Ontario schools from 1844 to 1876, a whopping thirty-two years! He was to become, as will be gleaned later in this essay, one of the most significant school promoters of the mid nineteenth century, and with others such as Malcolm Cameron, were forthright in the struggle for the revolution in the educational activities and systems. In the two individuals mentioned above, it was often a struggle of bitter opposition and conflict!

There still remain significant factors that need to be discussed and revealed here before delving into the meat of my argument, so to speak. The actual class teachings and settings had much to be desired. For instance, one, in most cases, poorly trained teacher would be crammed with numerous students of varying ages and abilities which made the task of teaching not merely challenging but a big fat joke! The problem of absenteeism and truancy did wreak appreciable setbacks and havoc in educating the populace of Ontario's children. In short, there was scant regard to schooling, and adults, parents, and children alike only paid attention to schooling after the domestic and field chores were exhausted and there was nothing else left to be done! Although this may appear somewhat exaggerated, it was, in most cases, a true and precise representation of the state of affairs up to the point of Reverend Egerton Ryerson. It is important to be reminded that parents and the local communities had quite a great say in the direction and form that education was taking then. It is also important to be reminded that these parents were themselves deficient in the educational experience. It was not until such reformers as Ryerson and John Strachan came on the scene with their new vision, that some of these responsibilities were slowly "wrenched" out of the palms of the hands of the disinterested. It, therefore, needs not the mind of a rocket scientist to envision the tremendous retardation in human development that would proliferate if these new classes of public school promoters had not stepped in through the weapon of legislation and unrelenting lobbying. To my mind, leaving the form and direction of education under the control of such disinterested, even ignorant Ontario citizens at the time, is analogous to allowing a "stoned drunken" man to operate a defective vehicle. Such a vehicle is bound to breed danger and cause wanton waste of human and other resources! And given the fact that most of the public school promoters of this era were of the view that formal education was crucial, as will be discussed later, their motives can be reasonably be assumed to be for the general good and based on their outlook of life, as was prevalent in the period being discussed.

It is to be said here and now as well, that this writer does not deem it relevant whether these public-school reformers were of the same ideology or not. In my humble view, it was not an important variable whether the promoters were engaged in a contrary, conflicting method to achieve their end. What is relevant, in my opinion, to the question under discussion here, is their goal, as can be gleaned from their deeds and words. In this respect, therefore, it is again, my opinion that both Malcolm Cameron and Egerton Ryerson, though on opposite sides of the fence in terms of approach on the education issue, had a similar and singular good of society.

There obtain a number of factors which make it compelling for one to regard the public-school promoters, of this period, as operating unselfishly. Let us consider the first case of Egerton Ryerson. It is quite tempting for his critics to chase to the conclusion that the gentleman was wantonly selfish and had a hot, fiery thirst only for power and accordingly, would do almost anything to stay in the limelight. In supporting their views, they may state, for instance, that he tried to gain more personal control of the education system to the expense of the local and ordinary citizens. But a close analysis of the situation would reveal that his action was selfless and unselfish. As Ryerson was convinced that firstly, education, formal education, was essential to the positive development of a people, he wanted to ensure that all got the opportunity to be thus exposed. Leaving the affairs of education matters to individuals not interested in it, would frustrate everything that Ryerson stood for or at least, that was his apparent thinking. Secondly, and closely related to the first reason, he was rightly convinced that ordinary individuals left to carry out the burden and responsibility of administering the process, may lead to stagnation of the people as they may not be sure of that which is worthwhile compared to that which is worthless!

It can easily be demonstrated that Ryerson, a leading school promoter, was motivated by a desire to advance the general good of society and not his own selfish class interest.

Ryerson is quoted in the text, *Unequal Union* (P238, Ryerson, Stanley B.) as follows: "That he looked upon the 'Rate Bill' Clause (taxing all property owners for school purposes, instead of parents only) above all others to be the poor man's clause and at the very foundation of a system of education. It was objected to by certain of the wealthy whose children attended private schools, one of such a Methodist—a magistrate, looks not beyond his own family. He says he does not wish to be compelled to educate all the "brats" in every neighbourhood, is the very objective of the clause of the bill; and in order to do so, it is proposed to compel selfish, rich men to do what they ought to do, but what they will not do voluntarily."

This quote, a long one by the way, spoke to and condemned any form of selfish aggrandizement in exchange for the benefit of the many, ordinary people. It is quite evident, from the above quote, that Ryerson resented selfish acts and was willing to work on behalf of the local people of Ontario.

Malcolm Cameron is to be considered among the public-school promoters and he also can be shown to have been working for societal interests and not to his selfish end. The very policies by the Cameron School Act of 1849 were living testimonies to this fact. The Cameron Act argues for the participation of the local people, as it was his view, though I disagree, that this was the way to promote the education of Ontarians during that period. Whether his view is agreeable or not, is not the point. However, what can be certainly eliminated from the Cameron's list of objectives for Ontario was that he was bent on satisfying a selfish end.

To quote Houston and Prentice (P126) in their text, *Schooling and Scholars in Nineteenth Century Ontario*, "What the Cameron School Act wanted, it would appear, was the greater local autonomy in education affairs."

Surely there is nothing selfish in Cameron or anyone, for that matter, wanting the local people to have greater control and decision-making opportunities in matters that concerned them. The only challenge here, in this recommendation, was his misread that the people, at that stage, had not the wherewithal to run a successful school system. However, by elaboration and extension, a general desire for local autonomy implies some kind of need to have local townships and communities have some semblance of control which he thought, at the time, was important. As I alluded earlier, I do not lend consensus to this view. This, without contradicting my point with Cameron earlier, proved that he had more than self-elevation in mind, in promoting his views and policies during this period. He had a conviction for the people, although it was an erroneous one!

There were yet other significant supports to the argument that public-school promoters, such as Ryerson, were acting for the society's general good. For example, by the School Act of 1871, the common school had become a public institution. That is to say, parents and guardians had no longer the responsibility to pay for their children's education and even more importantly, was the related fact that all funding, etc., fell under the portfolio of the provincial government. This School Act was largely as a result of these public-school promoters, especially, Ryerson. It is then, more than reasonable to infer, that if a set of parents did not have to pay for the education of their young ones, one would more likely be willing to make use of the system. This in itself, apart from revealing the great vision of such school promoters as Ryerson, did something even more significant, in my humble view. What is it that it did, you may be questioning at this point? The response is that the promoters, undoubtedly, reflected an unquestionable intent to better their communities on the whole. The intent, in this case, was to place education in the reach of ordinary folks who might otherwise, not be able to acquire any form of education due to pecuniary challenges. A second factor was that the same Act, alluded to in the aforementioned, had created a new category of, and a more advanced set of schools, other than the elementary ones. These schools were to be known as the high schools which are around up to this day. This, again, was with the view to making more Ontarians have the benefit of a more advanced and deeper and more meaningful, formal, educational training. It should be noted that a similar set of events was taking place all over North America as far as education was concerned.

A great philosophical view that evolved in the early nineteenth century and metamorphosed among school promoters during the century with respect to education, was another crucial factor in the interests of these scholars during the mid nineteenth century. Firstly, was the gradual and overwhelming view of the promoters that education could be utilized as a means of making the society a better, more tranquil place to live. It was believed and argued that in order to avoid or eliminate a dangerous or unsuitable environment, formal education was imperative. It was the perceptions of many such promoters that education could be used as a tool for the protection of the environment for all. It was out of this philosophical viewpoint that many public-school promoters' interests lay, not in any promotion of self but quite the contrary. It was to advance the wellbeing of the community on a whole and the growth, betterment, and improvement of all the local folks.

To quote Prentice in her book, *The School Promoters* (P25), "The movement to send all children to school was one to bring sanctity and order to human affairs, according to Ryerson."

The promoters were, by and large, acting on the basis of a new outlook, insight, if you will, that was pervading the minds of mid nineteenth century Ontarian and North American scholars and thinkers as a whole.

This new outlook and thinking which the new school promoters were involved in, to the cause of education, were reflected in numerous instances. It became a common thinking among North American philosophers and prominent others of this time, that humans' physical nature needed to be feared. To nullify this threat that homo sapiens presented to the society, there needed to be some kind of restraint, and this was exactly the purpose and objective of formal schooling. It is from this point of view, and no other, that the work and aspirations of these lobbyists for public schooling must be viewed. It is with such deep comprehension and insight that one will be made to understand why it is that Ryerson, for example, had to keep education under such tight control and "heavy manners" as he seemed to have been, especially in the eyes of his many critics. There was also the related view, around this time, that stemmed from the idea of the imperfection of man and I would hasten to state, women as well. It was the belief that the body and soul, whatever that is, must be united and this could only be achieved through a sound and thorough education. Hence again, an explanation of such aggressive acts with regard to the promotion of public schooling in this time, by public school promoters. It had not much, if anything, to do with selfish interests.

John Strachan one of the promoters/philosophers at the time had this to say, "The body and soul must be united to make the perfect man." (P27), Prentice in her text, *The School Promoters*.

Many other philosophical arguments and beliefs were a direct influence of the promoters of the day. There were school promoters and scholars such as Daniel Wilson who thought that starvation of the mind was significantly more of a criminal act than was the starvation of the stomach. With such serious views on the importance of the educating of its people, it should now be clear that the kind of efforts and fuss furnished in the mid nineteenth century about education can be explained quite reasonably and logically, from this standpoint. The great philosopher, psychologist, and historian, John Locke had a profound impact on the promoters' aggressive intent on ensuring that education be available for all to access, as well. His theory was that a child was born into the world as a tabula rasa. That is to say, that the mind of a newly born homo sapien was a blank slate ready to be influenced one way or the other. It was crucially important, according to this theory, that education reached this child before he or she be influenced by negative factors that could serve to undermine the general good of society which was the intent of the school promoters, by and large. Other scholars such as Henry Esson had similar views and were quite effective in influencing the promoters' drive for education.

The fact that the school promoters' actions were based on the need to advance the community's growth and progress can be gleaned in other ways as well. There was a need by a significant numbers of these school-promoters to reduce the overall economic costs of running the Province and their respective jurisdictions. This, in turn, would mean less taxation for the local people, which can be regarded as the benefit and progress of the society as a whole. This point is clearly supported when one examines such statements as was made by John Strachan, already mentioned earlier in this essay. He has been quoted in the Prentice's *The School Promoters* (P50) to have said: "Education would result in the comparative emptying of jails in the province and relieve the courts of a good portion of their business." This statement speaks clearly to the fact that the promoters had the broader society at heart in their struggle with educational reforms.

There was also the need to obliterate or reduce rampant materialistic outlook and to foster instead, an intelligent, decision-making populace. A group of individuals who would be able not merely to be engaged in meaningful, worthwhile activities and discussions but being competent and learned enough to assume political and administrative roles and ambitions. These were some other, deep, underlying reasons for the public-school promoters' involvement during this epoch, as can be gathered from reading the work of the great historian, Alison Prentice and many others.

In concluding, it is my forthright belief that the mid nineteenth century public school promoters' overwhelming goal was for the betterment of Ontario as a whole. This sudden enlightenment and thrust to the radical changes that Ontario and many other Western nations were to witness, arose predominantly as a direct consequence to the forthright influence of the philosophers of the time. Their theories and teachings were ones which promoted the education of the wider masses for the benefit of the society on a whole. As a matter of fact, according to the theories embraced by so many scholars of the time, education was the only hope to contain or curtail the annihilation of large groups of nations at the time. It was the only way to mould, in the minds of the young, desirable, sociable behaviour which would be beneficent for the proliferation of a tranquil and workable society. It was mainly out of this uncontrollable, yet powerful urge, stemmed and emanated out of the psychological impact of these school promoters, that to a significant extent, prompted their actions. From the readings available, and upon which I have come across, there is no significant evidence to conclude that the school promoters were selfish in their cause.

Notwithstanding that which has been stated above, and without any contradiction to my general position, events and phenomena are not always in black and white. To be more practical, there might have been a few individuals or certain activities that might have triggered some question as to the motive of certain of the acts of mid nineteenth century school promoters. Could it have been to the interest of self or the group or both? One such questionable action, and I am sure that they may be a few others, that could be argued by critics, wary of the public-school promoters, is somewhat reflected in this quote from the text, *Catholics Education in Politics in Upper Canada*, authored by Franklin Walker. He wrote (P65), "Ryerson was insistent that all pupils be exposed to the Christian teachings . . . he was especially concerned that an attempt was being made to establish a system of Godless education."

Notwithstanding the claim that an isolated act by a promoter may be inclined to selfish end, it does not necessarily make this school promoter's or any other such promoters' acts in the general sense, selfish. For example, it is an unquestionable fact that to be an alcoholic, one has to consume alcohol, but it would be wantonly absurd to accuse an upright individual who takes one or two drinks of alcohol occasionally but never get drunk, or be any threat to his family's or society's wellbeing, or himself, to be classified and categorized as an alcoholic! The act of imbibing alcohol is not what counts in categorizing such an individual. It is the amount and regularity with which the individual consumes it. So it is, in the case of the public-school promoters in mid nineteenth century. A few questionable actions do not change the fact that these individuals were, by and large, and in significant numbers, working in the interest of the larger society.

With all that said, and in conclusion, I reiterate, that the school promoters, in general, were unselfish in their duties to education. That to a large extent, their interests and deeds, with respect to education, were motivated by a driving desire to the advancement of the general good of society and not by their own selfish class interests as the one-sided, probable, mentally myopic critics would have us believe. Having said it all, I rest my case!

TEST YOUR LEARNING

---

(1) Using information from this essay, additional resources such as your local and/or school library, (a) explain whether the author's presentation adequately and correctly reflects the views he proffered that the local school promoters of mid nineteenth century were not acting on selfish motives. (b) Were you impressed, or not, in the way he lends support to his arguments? Explain.

(2) (a) Would you agree or not that the author has given sufficient facts to argue his points. (b) Were these solid points? Explain.

(3) Using information from the essay, additional resources such as your local and/or school library or by interviewing an expert historian, (a) did the author adequately and convincingly present his arguments in support of his views? (b) Do you think his points were mostly original? (c) Could the premise of his arguments stand up to critical scrutiny? Explain in each case.

(4) On a scale of 1 to 10, 10 being the best, 1 the worst, (a) how would you rate this essay?

(b) How would you rate his introduction to the essay, using the same scale? Explain. (c) Are there any supporting details that the author provides, that were not relevant to the discussion or in making his point? (d) Could he have provided more supporting details to strengthen his arguments in the presentations of his points? If yes, what arguments or points could he have included? (e) Could he have supplied more possible contrary points of view to show balance in his presented arguments?

(5) (a) Did the essay change your opinion or any perception(s) you have/had with regards to the origin and introduction of formal education in the West? If so, in what way did it change your view on the subject? If it did not change your opinion on the matter, explain why not. (b) Did you learn new information or facts from the reading? Explain.

BIBLIOGRAPHY

Adams, Howard. *The Education of Canadians, 1800-1867: The Roots of Separatism*. Harvest House, 1968.

Beidelman, T. *A Matrilineal People of East Africa*. Holt, Rinchart and Winston Inc.

Cragg, Wesley. *Contemporary Moral Issues*. McGraw Hill, Ryerson Limited, 1992.

Derv, Naranjo-Hueb. *Profile Feminism Yesterday and Today* (Second Edition).

Feinberg, Joel and Gross, Hyman. *Philosophy of Law* (5th Edition). Wadsworth Publishing Company, 1995.

Guerrine, Anita. *Experimenting with Humans and Animals*: From Galen to Animal Rights. Baltimore: John Hopkins University Press, 2001.

Houston, Susan E. and Prentice, Alison. *Schooling and Scholars in Nineteenth Century Ontario*. Toronto: University of Toronto Press, 1988.

Judy, Robert. *A History of the African People*. Hunter College and Graduate Centre, Charles Schribner's Sons.

Larquin, Paul F. *High Tech Harvest: Understanding Genetically Modified Food Plants.* New York, Columbia University Press.

Lewis, Jone Johnson. *Abortion History: A History of Abortion in the United States*. Mackinnon, Barbara. Human Cloning: Science Ethics and Public Policy. Urbana: University of Illinois Press, 2000.

McGee, Glen. *The Human Cloning Debate*. Berkeley, Ca: Berkeley Hills Books, 2002.

Mandani, Mahmood. *Citizen and Subject: Introduction*. Princeton University.

Mhlaba, Luke. *Local Culture and Development in Zimbabwe: The Case of Matabeleland*. Scandinavian Institute of African Studies, P.O. Box 17035-751 Uppsala, Sweden.

Mlama, Penina M. *Culture and Development: The Popular Theatre Approach in Africa.* The Scandinavian Institute of African Studies, P.O. Box 17035-751 Uppsala, Sweden.

Prentice, Alison. *The School Promoters*. Toronto: McClelland and Stewart Limited, 1977.

Ryerson, Stanley B. *Unequal Union*. Toronto: Progressive Books, 1988.

Shiva, Vandana. *Tomorrow's Biodiversity*. London: Thomas 2000.

Silver, Lee M. *Remaking Eden: Cloning and Beyond in a Brave New World*. New York: Avon Books, 1997.

Tudge, Colin. *The Impact of the Gene From Mendel Peas to Designer Babies*. New York: Hill and Wang 2000

Turnbull, Colin. *Tradition and Change in African Tribal Life*. The World Publishing Company.

Uschan, Michael V. *Heroes and Villain*—Martin Luther King Jr. Lucent Books, 2004. 27500 Drake Road, Farmington Hills, M148331.

Walker, Franklin A. *Catholic Education and Politics in Upper Canada*. Canada: J.M. Dent and Sons, 1955.

Walters, Brent and Cole-Turner, Ronald. *God and Embryo: Religious Voices on Stem Cells and Cloning*. Washington DC Georgetown University Press, 2003.

Washington, James M (Edited by). *I Have a Dream: Writings and Speeches that Changed the World*. HarperCollins Publishers, 1992.